LAKES

LAKES

THEIR BIRTH, LIFE, AND DEATH

JOHN RICHARD SAYLOR

TIMBER PRESS
Portland, Oregon

Cover art by Florida LAKEWATCH (lakewatch.ifas.ufl.edu)

Published in 2022 by Timber Press, Inc.

The Haseltine Building
133 S.W. Second Avenue, Suite 450
Portland, Oregon 97204-3527
timberpress.com

Printed in the United States of America
on paper from responsible sources

Text design by Sarah Crumb
Jacket design by Hillary Caudle

ISBN 978-1-64326-048-8

Catalog records for this book are available from
the Library of Congress and the British Library.

To Amy

CONTENTS

INTRODUCTION

Pick a lake, any lake. It could be the one you are sitting next to right now or the one you visited last summer. Or it could be a random lake chosen by simply pointing your finger at a map. Whatever lake you choose, large or small, near or far, two things about it are true. First, it didn't used to be there, and second, someday it will be gone.

Most lakes are born of violence. Be it the slow-motion violence of glaciers, the most prodigious of lake-builders, or the considerably more rapid violence of volcanoes, earthquakes, or (rarely) meteor impacts, most of these water-filled depressions that we call lakes become a part of our planet's topography via the application of incredible force.

Like us, lakes are born, live, and then die. Like us, some lakes seem to have ordinary lives, while the lives of others are considerably more exotic. And much like the anguish, joy, and tumult that exist in even the most mundane of human lives, even the most ordinary of lakes experience a kind of scientific drama not unlike the emotional drama of their human counterparts. In spite of the calm pastoral nature we see in the small duck pond down the street or the larger lake with the sailboat gliding across it in the distance, these everyday lakes experience an array of profound and wondrous processes, often hidden in plain sight.

The phenomena existing in lakes are as fascinating as they are commonplace. From the surface tension that enables water striders to walk on water to the miraculous characteristics of the water molecule that allows ice to float on water (and in turn allows *us* to walk on water), even the most mundane of lakes are filled with magic. Lakes experience the magic of evaporation, this silent, invisible process that can steal a half-dozen feet of lake water in a year. Lakes are intimately involved in the consumption, production, and transfer of oxygen and carbon dioxide, which so much of life depends on. And lakes often experience overturning, a process critical to the life within, where water from the lake depths suddenly rises up, replacing surface water that, in turn, descends to the lake bottom.

Then, there are the more exotic lakes, those with rock-star lives. There is Lake Nyos in Cameroon, a lake that accrued dissolved carbon dioxide in its depths for untold years, and then exploded in 1986, releasing about a quarter of a cubic mile of gas and killing over a thousand people in the process. There are lakes that lay beneath a sheet of ice a mile or more thick in the Antarctic, at least one of which is as big as Lake Ontario, and so isolated that the organisms living there have not been exposed to the atmosphere for over 100,000 years, a time when mastodons still roamed the earth. And there are the Carolina bays, a series of over 80,000 lakes located in the Atlantic Coastal Plain of the United States, all of them perfectly elliptical, for some unknown reason, all of them oriented in the same direction, as if pointing to something important.

All of these lakes, exotic or pedestrian, were at one point created, and at some point in time they will all disappear. Some lakes are formed by glaciers, others by volcanoes, and some, notably the Carolina bays mentioned above, were formed by processes that are still not understood. All will one day be gone. Many will experience a slow senescence as silt and soil carried to them by the very streams and rivers that sustain them, settles and slowly fills them, transforming these lakes from deep blue bodies of water to shallow marshy expanses, to swamps, and finally to nothing more than a green meadow, leaving behind little hint of their former existence. Some lakes will take millions of years to fade away. Others will die within our lifetime, often due to human activity, much like the Aral Sea, once one of the largest lakes on Earth and now almost completely gone, devastated by the diversion of its water sources for irrigation and other human needs.

This book is about the physical phenomena that occur in lakes: the phenomena that cause lakes to form as well as those that cause lakes to die. We'll look at all of the interesting phenomena that occur throughout the life of a lake, from those lakes that last only for weeks, like lakes formed by landslide dams, to those that have existed for millions of years, like Lake Baikal. We will explore lakes from the most ordinary to the most extraordinary, from the mundane to the magical, from the poles to the equator. All kinds of things happen in lakes, including your favorite lake, and maybe the lake that you are sitting next to right now.

Let's jump in.

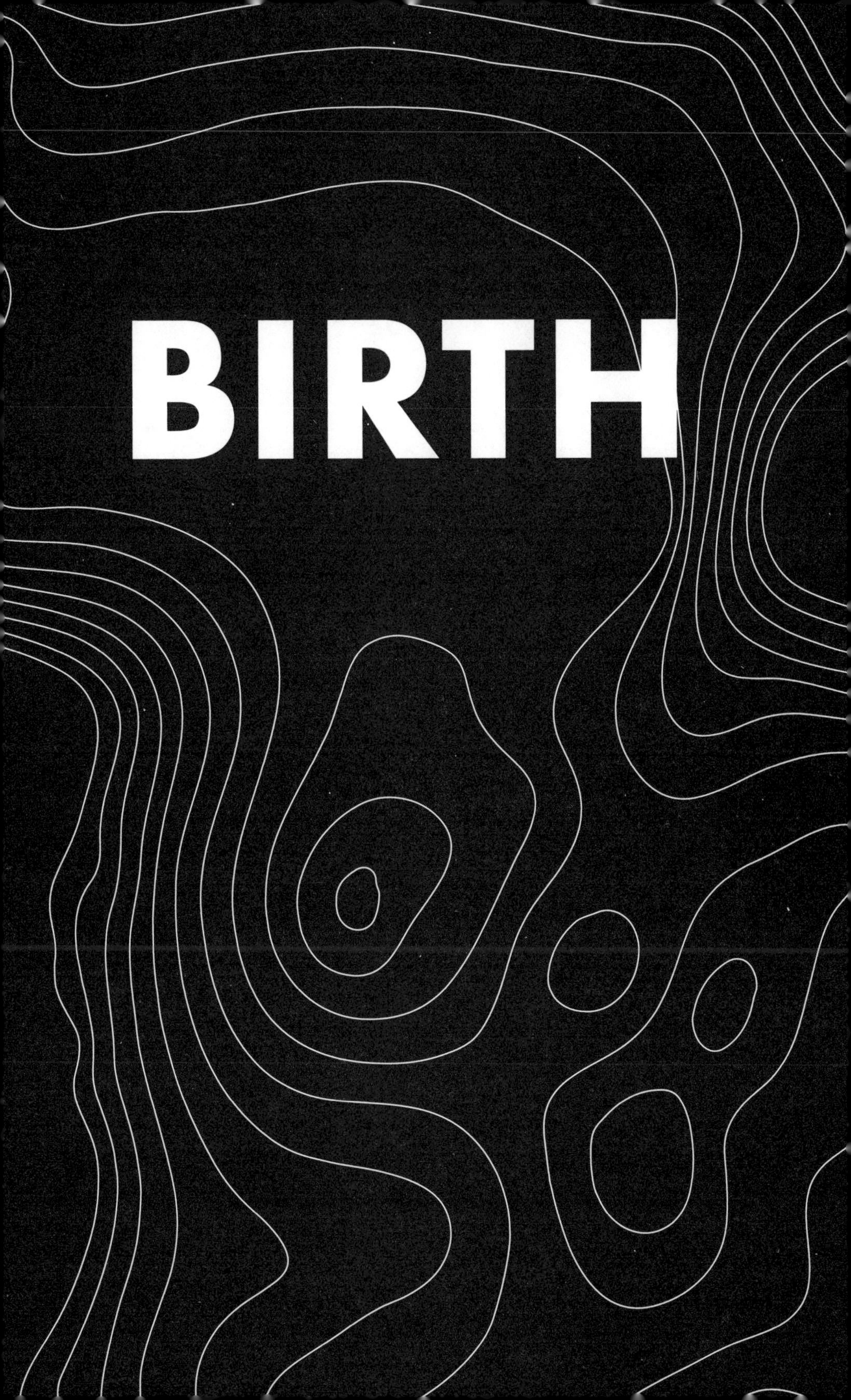

BIRTH

1. GLACIERS

The Master Creator

It may not seem that glaciers have a lot to do with lakes. Today these masses of ice and snow are found near the poles or high up in the tallest of mountains, far from the lakes one finds in more temperate latitudes and altitudes. But glaciers once covered an enormous part of the planet, in the Northern Hemisphere covering all of what is now Canada and portions of the United States, extending as far south as Indiana. They covered most of northern Europe and the northern regions of much of Asia. In the Southern Hemisphere, an ice sheet extended outward from the current boundaries of Antarctica all the way to the edge of the Antarctic continental shelf. And, wherever glaciers spread over land, they left prodigious quantities of lakes behind.

To have a lake, you need some kind of depression in the ground, a basin with a rim on all sides high enough to hold significant quantities of water. This doesn't happen as naturally as one might think. When mountains form, they tend to create terrain that tilts more or less uniformly downhill from peak to foot, a situation that typically does not result in the formation of a lake basin. Tectonic plates may warp, tilt, or fracture to form basins that ultimately become lakes. But this results in far fewer lakes than those made by glaciers.

Lakes can form in very flat regions, like the Great Plains. In such regions, rock beds made of limestone or other types of soluble rock can be acted upon over time to form lakes via a process called solution. Here, a small, shallow dip can become exaggerated as accumulated water becomes enriched in carbon, which in turn forms carbonic acid that can dissolve rocks such as limestone. The surface slumps as the underlying rock dissolves, resulting in a lake. Excellent examples can be found along Croatia's Dalmatian coast and in Florida. Lakes can also be formed by volcanoes and by landslides that dam up a river valley. But, all of these types of lakes—solution lakes, lakes formed by various tectonic processes—all of them take second place to glaciers when it comes to number of lakes formed.

Most of the planet's lakes were formed by glaciers. While estimates of the exact percentage of lakes formed in this way vary, these estimates always exceed 50 percent and range to as high as 90 percent. Thus the number of glacially formed lakes is very large simply because the total number of lakes found on Earth is enormous.

Precisely how many lakes exist is difficult to say for several reasons, not the least of which is the difficulty in agreeing upon a lower size limit for what exactly constitutes a lake. The introductions of limnology (the study of lakes) textbooks often include a good deal of fretting over the difference between a pond and a lake. But if, for the sake of simplicity, we include ponds in our inventory of lakes, matters are little helped since now the question becomes what is the lower size limit for a pond? Ultimately the decision is arbitrary and perhaps it is best to just count the number of lakes larger than a particular size. In the first chapter of *The Lakes Handbook*, C. S. Reynolds does precisely this, noting that if one limits the definition of a lake to a freshwater body having an area of more than 24.7 acres, existing data suggest there are 1.25 million lakes on the planet. If one then adds those bodies of water having an area between 2.47 acres and 24.7 acres, the number increases by 7.2 million, giving at least 8.45 million lakes. To put these numbers in perspective, note that a perfectly circular lake having an area 2.47 acres has a diameter of 370 feet, and a 24.7-acre circular lake has a diameter of 1171 feet. The estimates of total number of lakes only get larger if we include yet-smaller bodies of water, manmade lakes, or ephemeral lakes (ones filled during only certain times of year). In any event, no matter how the counting is done, our planet is well-endowed with lakes. This is a good thing, given how many species rely on them to survive, our own species being no exception.

Eight million is a lot of lakes, and if it weren't for glaciers, that number would be far smaller. Glaciers are prodigious lake builders. This is an odd thing to ponder, particularly if you happen to be vacationing by a lake right now, perhaps during the summer, cooling yourself as you float upon the water surface. Perhaps you are lazing by one of the Finger Lakes in New York State, or perhaps one of the many lakes found in Minnesota, or maybe you are touring the scenic English Lake District. These lakes differ in many respects, not the least of which is their location, but they all share the same origin story—they were all formed or modified by glaciers, these mammoth sheets of ice. This thought does not naturally come to mind as you laze by the lake shore during the heat of summer.

Glaciers create lakes in a multitude of ways. Sometimes they dump ground-up rocks and silt at the end of a valley (often a mind-boggling amount of ground-up rocks and silt), damming up a river valley and creating a long thin lake; the Finger Lakes are excellent examples. Other times they may grind away at pre-existing faults or cracks in the bedrock, gouging out a basin. Other times they shatter rocks via frost-thaw cycles, then slowly, over the millennia, drag the rock away, leaving a bowl behind.

Filling the created bowl with water is easily done, and often provided by more than one source. When they recede, glaciers leave massive quantities of meltwater behind, readily filling any dips or depressions in the moraine they also leave behind. Enormous quantities of water are also left behind in underground aquifers. Rainwater and streams or rivers entering a basin also supply a lake with water.

Although our interest here is in lakes, it is important to note that in many parts of the world, the very topography of the land is the way it is because of a glacier. While volcanoes and the motion of tectonic plates may be the creators of the raw hills, mountains, and valleys covering our planet, glaciers modify these features in powerful ways, creating the topography we have today. There are few things on Earth that have the landscape-sculpting power possessed by these miles-thick sheets of ice.

To understand how glaciers form lakes (and are forming lakes), let us pause to explore what glaciers are and how they work. A glacier is, essentially, a long flowing river of ice whose source is located in an elevated area where significant snowfall occurs, and whose termination point is a lower area where snowfall is relatively sparse. In the elevated region, snow accumulates, compressing older layers of snow into ice. This is a slow process, as are most processes related to glaciers, the layers of snow growing year by year, decade by decade, century by century. This upper region is called the accumulation zone, and lacking other processes, the accumulation zone would continually thicken.

In lower areas where snowfall is sparse, a process called ablation thins the glacier. Ablation is the combined action of evaporation and sublimation (the direct conversion of solid water into water vapor). This zone is referred to as the ablation zone and, lacking any connection with the accumulation zone, the ablation zone would quickly disappear. So, lacking a key piece of physics, ice in the accumulation zone would grow without bound, and ice in the ablation zone would quickly dwindle away to nothing.

But there is a key process preventing both of these scenarios from happening, a process perhaps the most difficult to wrap one's head around when thinking about glaciers—the simple fact that ice flows. The pressure of the growing layers of ice in the accumulation zone is simply enormous, causing the ice to flow downhill into the ablation zone, where it more or less replaces the mass of water lost via ablation. "More or less," because glaciers can advance and retreat. When snowfall in the accumulation zone exceeds the mass of water lost in the ablation zone, glaciers advance, the ice flowing forward and extending the tip (called the snout) of the glacier. When the mass of water lost from the ablation zone exceeds snowfall in the accumulation zone, then glaciers retreat, the snout moving upstream.

None of this occurs quickly, but the flow of glaciers is inexorable. It is a kind of slow-motion violence capable of obliterating anything in its path. And it is this flow of ice that is critical to many of the ways that glaciers create lakes.

The confusion about how ice would be worthy of the word *flow* and all its liquidy connotations is that, simply stated, it is not true that solids do not flow. To be sure, under the forces that something like an I-beam experiences, flow (at least in any observable sense) does not occur. But, when forces become sufficiently large, even metals will flow, as is the case when, for example, a copper bullet impacts steel, the relatively soft copper flowing and deforming under the very large forces at play during the brief period of the bullet impact event. The same occurs in glaciers, the ice slowly flowing downhill under the enormous pressures exerted by an ice sheet up to a mile thick.

This type of glacial flow dominates what are referred to as cold-based glaciers, glaciers whose temperature at the interface between the ice and the ground beneath is below freezing. Here, the glacier is frozen to the ground, with little sliding at the base. The bulk of the glacier moves because it deforms and flows downhill even as the ice wedded to the earth at its base moves very little. However, the real violence glaciers do to the ground occurs when they are (or become) warm-based. The temperature at the bottom of a "warm-based" glacier is above freezing, causing a film of water to exist at the ice/rock interface. This film of water permits the glacier to slide over the ground. The pressures beneath are still large, and the motion of the ice over the rock creates a layer of partially ground-up silt, pebbles, rocks, and even boulders. This debris then serves as a kind of grit, like rough-textured sandpaper—but sandpaper under enormous pressure. Warm-based glacial motion scores, scrapes, and grinds the bedrock, leaving all of the telltale signs of glaciation.

Though glaciers move quite slowly, there is a surprisingly large range in their velocities. Particularly fast-moving portions called "ice streams" can move many hundreds of feet in a year, while the more sluggish portions may travel only a few feet per year. In some locations, a glacier can be essentially static and unmoving for long periods of time. But regardless of their speed, flowing glaciers pick up and push aside a massive amount of material. Dust and debris from the surrounding area fall upon the surface of the glacier and are incorporated into the layers of snow where they are eventually integrated into the icy body of the glacier. Rocks from the size of pebbles to enormous boulders are plucked up by glaciers and transported downstream, in the direction of the ice flow. All of this material is referred to as moraine or glacial till, and ends up somewhere once a glacier retreats. The amounts of moraine can be enormous, easily dominating a landscape. All of this rocky material can dam up water flows, or simply leave a huge layer of moraine that itself has depressions in it capable of becoming lakes.

Broadly stated, glaciers form lakes via two processes. The first is the cracking, destruction, and removal of rocks, a process that typically occurs in the upper regions of the glacier, often very close to its origin near a mountain peak. This can form, for example, cirque lakes (described below), where a bowl is gouged out by the glacier and the resulting material deposited on the downstream side of the bowl, forming a basin that becomes a lake. The second process of lake formation concerns the ultimate location of all of the material that glaciers gouge out—the moraine and glacial till—the rock, soil, silt and other detritus glaciers pick up and move about.

In the popular imagination, glaciers are often considered to be pristine tongues of blue ice, and parts of glaciers do indeed look this way. But a glacier can also be fantastically dirty, filled with all manner of material that existed in its path. At the large end, boulders as large as a house can be moved by glaciers, sometimes plucked up at the base of the glacier and moved downstream, then left behind, sometimes in the middle of a valley where no similar rock can be found nearby (these are called dropstones or glacial erratics). At the other end of the size scale, glaciers can grind rocks up to the point where all that is left is an extremely fine powder called rock flour. Rock flour is easily transported by the melted water of a glacier and gives the meltwater an odd color, a kind of milky turbidity that has been variously described as opalescent, bluish, or greenish-yellow. These colors are used by glaciologists as a telltale sign that the flow of water into a lake or bay has originated as glacier meltwater somewhere in the mountains above.

"Moraine" gets pushed either to the side, forming lateral moraines, or in front of the glacier, resulting, ultimately, in a terminal moraine. Just like an enormous bulldozer pushing earth about, glaciers can dam up valleys and leave undulating quantities of moraine. All of these can fill with water, as described earlier, and form lakes.

One type of glacially formed lake, particularly common in the northern United States and Canada, is the kettle lake. There are countless numbers of these quaintly named lakes. Perhaps the most famous is Walden Pond in Massachusetts, whose shores were graced by Ralph Waldo Emerson and Henry David Thoreau, whose book, *Walden* is often seen as the origin of environmentalism. These lakes are formed not from the relatively simpler process where terminal moraine is dumped at the end of a valley, thereby damming up a river, but from a decidedly messier and in fact incompletely understood process. When glaciers recede, it isn't necessarily as one might imagine, the snout melting as a sort of uniformly receding line. In some cases, enormous chunks of ice break off as the glacier melts, staying behind perhaps for as long as a hundred years or more. Meanwhile the remainder of the glacier, the main tongue of ice, continues to recede, pulling back to a higher altitude or a more northerly location. In the simplest of cases, these blocks of ice are surrounded by moraine. Hence, as the ice block melts, the moraine serves as the bank of a newly formed lake, the water conveniently provided by the melted ice.

But the process can be more complex when the large blocks of ice aren't simply surrounded by moraine, but are also covered by it and contain it. A bowl shape will still result when the ice fully melts, but the resulting shoreline and lake bottom can have just about any shape. The process is poorly understood, but certainly is complex, as the slowly melting ice fills a basin formed by the surrounding moraine that is certain to adjust as the ice melts. Moraine once supported by underlying ice is bound to fall and shift. Moreover, weathering of the moraine once the ice is gone will further change the shape of the lake banks. Accordingly, kettle lakes exhibit a broad range of shapes, from the smooth and rounded to the highly irregular.

A glacially formed lake common to alpine regions is the cirque lake, also referred to as a tarn. These lakes are most commonly located close to the origin of alpine glaciers and are found in every mountain range that has experienced glaciation or is currently glaciated. In these headland regions, the rock of the mountain immediately adjacent to the head of the glacier experiences large variations in temperature. The glacier itself cools the region, but during the day sunshine increases the temperature to above freezing at the upper edge, melting some of the ice. The resulting meltwater, as well as precipitation and snowmelt from higher up on the peak, flows into cracks in the rock near the glacier edge. At night, temperatures drop below freezing, and the water in the cracks freezes and expands.

Repetitions of this frost-thaw cycle shatter the rock, causing rock falls and exposing other cracks. Since this process occurs primarily at the head of the alpine glacier where the topography is already steep, these frost-thaw cycles tend to form vertical cliffs along the headland of the glacier. The pieces of shattered rock eventually make their way beneath the glacier, which uses these bits of rock to slowly grind out a curved bottom as the glacier moves downstream. The resulting glacial till continues to move downstream, and some is deposited on the far side of the cirque, creating a lip that encloses the basin. Upon glacial recession, what remains is a more or less circular lake with a steep vertical headwall on the side nearest the mountain peak and a lip composed of till on the side opposite the headwall. Cirques often resemble amphitheaters or look like giant armchairs in the upper mountain valley. Fed by runoff from precipitation and snow melt, cirques are the gorgeous mountain lakes often seen near peaks.

While cirques are usually found near mountain tops, they can also be found in lower regions of the mountain, and when this happens, they may appear as a line of round lakes distributed along the mountain valley and connected by a river, the effluent of one cirque feeding the cirque beneath it. Such lake chains are sometimes referred to as paternoster lakes, the name derived from the similarity of this chain of roughly circular lakes to rosary beads, paternoster being Latin for "Our Father," the first prayer in the rosary. A striking example of such a lake chain can be found along the Swiftcurrent Valley in Montana's Glacier National Park.

This description of how glaciers grind and scrape rocks, accumulating all manner of mud and silt, gives an impression of glaciers as messy things. And, indeed, the interface between the sole of the glacier and the ground can certainly be a very dirty place. Moreover, the surface of glaciers can be soiled with bands of dirt and speckled bits of moraine. But it is also true that glaciers can be extraordinarily beautiful. When one reads textbooks on glaciers, in between the terminology-filled pages and the descriptions of moraine and rock scrapes, are also lyrical passages describing the stunning beauty of glaciers. It seems these scientists, these glaciologists, could only contain their joy and wonder for so long and just had to wax poetic for a time, before returning to their scientific selves. This is nowhere more apparent than in the book *Glaciers of North America*, written in 1897 by Israel C. Russell, professor of geology at the University of Michigan. In describing crevasses, the enormous cracks in the ice that open up when glaciers move over uneven terrain, he writes:

> *The sides of crevasses are frequently hung with icicles, forming rank on rank of glittering pendants, and fretted and embossed in the most beautiful manner with snow wreaths, and partially roofed with curtain-like cornices of snow. These details are wrought in silvery white, or in innumerable shades of blue with suggestions of emerald tints. When the sunlight enters the great chasms, their walls seem encrusted with iridescent jewels. The still waters with which many of the gulfs are partially filled, reflect every detail of their crystal walls and make their depth seem infinite. No dream of fairy caverns ever exceeded the beauty of these mysterious crypts of the vast cathedral-like amphitheatres of the silent mountains.*

Perhaps it is this glacial beauty and magic that imparts itself upon the equally beautiful lakes formed by glaciers.

Another thing one notices when reading books and journal articles on glaciers and the lakes they form, is the focus on the Northern Hemisphere. Regions like Canada, the northern United States, Greenland, Iceland, and northern Europe get virtually all the attention. Very rarely does the literature discuss South America or Africa or Australia. Shouldn't there be glacially generated lakes in these regions, formed perhaps by ice that extended from Antarctica? The answer to this question is, mostly, "no" simply because, to have a lake, you must have land, preferably land near a pole. In the Southern Hemisphere there isn't a whole lot of land near the pole, at least compared to the Northern Hemisphere.

When an ice age begins, glaciers advance in two ways. First, they originate on the peaks of mountains and advance downward, a process that happens even in mountain ranges located on the equator. The second way glaciers advance is from the poles toward the equator, a process that forms enormous continent-spanning glaciers. So, to have a very large glacier and hence a lot of glacially formed lakes, it helps to have a lot of land near the poles. With this rubric in mind, let us compare the

Southern and Northern Hemispheres by observing just how much land exists between an arbitrarily chosen latitude of 45 degrees and the pole at 90 degrees. This exercise yields a surprising result. In the north, there is a lot to talk about. Virtually all of Canada exists north of 45 degrees north—the entire nation except for parts of Nova Scotia, a bit of southern Ontario, and a very small bit of New Brunswick. In the United States all of Alaska, Washington, and North Dakota are north of this boundary along with most of Montana, and significant parts of Minnesota, South Dakota, Wisconsin, Michigan, and Maine. Virtually all of Russia is north of 45 degrees along with most of France, all of Germany, Poland, Hungary, Ukraine, Denmark, the United Kingdom as well as the nations of Norway, Sweden, and Finland (and of course Iceland and Greenland). That adds up to a lot of land.

If we now turn our attention to the Southern Hemisphere, what we will see (and if someone hasn't pointed this out to you before, this may come as a bit of a surprise) is almost nothing. South of 45 degrees exists the southern tip of New Zealand, the Falkland Islands, and a portion of the southern tip of South America—a bit of Chile and Argentina. Save for a few small island outposts such as the South Sandwich Islands or Heard Island, that's all there is. None of Africa is south of 45 degrees. None of mainland Australia is south of 45 degrees. Indeed, if you were to take a globe and look down directly over the South Pole, your primary impression would be one of oceans. If you do the same by staring down at the North Pole, your primary impression is one of land. In short, the prodigious covering of the Northern Hemisphere with glacially produced lakes exists because that is where the land is.

If you are any kind of a student of geography, you have probably noticed that the above description of the Southern Hemisphere contains a glaring omission. What the Southern Hemisphere does have between its pole and 45 degrees latitude is something the Northern Hemisphere doesn't—an entire continent: Antarctica. Furthermore, Antarctica is enormous and nothing if not glacially covered. But, this large continent is not a place where glaciers *used* to be—it is an intensely glaciated place *now*, with almost no exposed land. There hasn't been the process of glacial recession and the concomitant lake formation that exists in the Northern Hemisphere. Once again, when we talk about lakes and particularly glacially formed lakes, the discussion mainly concerns the Northern hemisphere. But, there is an interesting exception.

Although Antarctica does not have lakes in the typical sense, which is to say bodies of water with a floor composed of rock or soil and a surface exposed to the atmosphere—it does have many lakes of a very different kind. Termed subglacial lakes, Antarctica's lakes are pods of water located at the boundary between the bedrock of Antarctica and the ice sheet lying above. Some of these lakes are enormous—at least one, Lake Vostok, is as large as Lake Ontario. One might argue that these lakes are somewhat like a typical lake in winter with a sheet of ice on its surface. However, this sheet of ice can be miles thick and does not melt away in the summer. These lakes contain water that has not been exposed to the atmosphere in hundreds of thousands of years or, according to some estimates, for millions of years. Covered as they are in thick ice, the waters in these lakes are utterly dark. What exists or might exist in the black depths of these mysterious lakes deep beneath the glacial surface is a subject we will return to in a later chapter. Suffice it to say that in the cold darkness of these lakes exist life forms we are only just learning about.

The above has hopefully demonstrated the importance of glaciers in forming lakes, a process that occurred when glaciers dominated those parts of our planet that, today, experience ice only in the winter, if at all. But, while the importance of these glaciers on the formation of lakes might seem to be one of historical significance, it should be noted that the process is ongoing. Indeed, in the currently popular discussion of global warming, the fact is often lost that Earth is currently in the midst of an ice age, specifically the Cenozoic ice age, which began approximately 37 million years ago. That may seem like a long time but in geological terms, it is relatively brief. For context, note that the ice age prior to the Cenozoic, the Late Paleozoic, began some 300 million years ago.

Ice ages exhibit periods of glaciation followed by interglacial periods, time intervals when glaciers recede and the planet experiences warmer conditions away from the poles and mountain peaks. There is much debate about how long these periods lasted (and will last). There is, however, general agreement that during an ice age, glacial stages last for about 100,000 years and interglacial periods last about 20,000 years. The peak of the last glacial period is thought to have been about 18,000 years ago, and it is this period of time that colloquially is often referred to as the ice age. But it must be noted that our ice age, the Cenozoic, continues.

And, global-warming concerns notwithstanding, our current interglacial period is actually not as warm as prior ones have been. The peak of the last interglacial period was about 125,000 years ago and was a time when flora and fauna were found much farther north than today, as evidenced by fossils of beetles and trees in locations such as Baffin Island. The amount of ice on the planet during this last interglacial period was much smaller and world sea levels are estimated to have been 16 to 26 feet higher than today.

So, while it is wise to take note of the potential impact humans can have on the planet in terms of its climate, we should also not lose sight of the fact that our ice age is far from over. Ice, and the process of glaciation, is still going strong. Indeed, there is currently so much ice covering Earth that most of our freshwater exists not in lakes, rivers, and streams, but in glaciers where as much as 75 percent of the planet's freshwater is held. Currently there are 400,000 glaciers and ice caps on the planet, covering approximately 11 percent of the world's land mass. Another 14 percent of the world's land mass has frozen ground for all or part of the year. When it is winter in the Northern Hemisphere, more than half of the land on Earth is covered by snow. Nor is just land affected; when it is winter in the southern hemisphere, as much as 7.7 million square miles of the Southern Ocean will be covered in sea ice. The process of glaciation, with all of its attendant grinding motion and deposition of rock and silt, is occurring right now. The slow inexorable formation of lakes continues. It will continue until the present ice age comes to an end and then not again until the next one arrives, destroying old lakes, and creating new ones, millions of years in the future.

2. VOLCANIC LAKES

An Inherited Violence

In contrast to the slow-motion violence that forms glacial lakes, volcanic lakes are created by decidedly faster processes—processes literally explosive in nature. Of course the lakes that can be found in the various depressions that volcanoes create are formed long after all the Sturm und Drang, and at a much slower pace than the volcanic eruption itself. Nevertheless, an understanding of the lakes that originate in the craters and calderas and behind the lava dams formed by volcanic activity depends on a general understanding of how volcanoes work in the first place. So, we would do well to begin by exploring these violent spewers of magma and ash.

We are all aware of the existence of volcanoes, and lest we forget, we are periodically reminded of their bad behavior by news of a volcano erupting in Hawaii or Indonesia or Iceland. We see video of orange lava flowing into the sea and local residents fleeing towns and villages by any means necessary. But as devastating as recent volcanic eruptions have been, to get a real sense of what a volcano can do, we need to go back a bit in the historical record, which provides ample evidence of the truly destructive power of a volcanic eruption.

Perhaps the best-known example of a killer volcano is the AD 79 eruption of Vesuvius. The event buried the cities of Pompeii and Herculaneum and killed an estimated 16,000 people. Incredibly, we know some of the specific details of this eruption in part from letters written by Pliny the Younger, whose uncle Pliny the Elder perished while seeking to explore the eruption as well as to rescue those trapped at the base of the mountain. Pompeii was completely covered in ash and volcanic debris, the devastated city not rediscovered for over a thousand years. Extracting the story of the eruption was facilitated by the fact that the hardened volcanic debris left cavities where the encased bodies of Vesuvius's victims decomposed. These cavities have been exploited by archaeologists who filled the spaces with plaster to create three-dimensional replicas of the dead frozen in their death throes. These statues reveal how quickly a volcano can kill, showing people crouched down, attempting to shelter their children, or grasping bags of silver or gold coins.

But as bad as Vesuvius was, volcanoes can be even more destructive. In 1815, Mount Tambora, a volcano in Indonesia, exploded and sent an enormous quantity of rock into the air. The strength of the explosion is illustrated by the distance these rocks were ejected, some as far as 25 miles away. An estimated 20 cubic miles of ash was sent into the atmosphere during the blast. The repercussions were global; so much of the sun's light was blocked by the ash that 1816 was sometimes called "the year without a summer," and harvests failed in Europe and North America. The blast was heard over a thousand miles away, and 10,000 people were killed. In the ensuing months, another 82,000 people died on nearby islands due to famine and disease.

Even more destructive, though with a lower death toll, was the eruption of Krakatau in 1883, which literally impacted the entire planet. It is thought that the Krakatau eruption was the loudest sound in history. Heard thousands of miles away, the shock wave was detected by barographs around the world. The eruption caused the deaths of 36,000 people in the surrounding area (most from the tsunami that followed the eruption), and like Tambora, impacted the entire planet via the sheer volume of material it ejected into the atmosphere. When Krakatau exploded, an entire island was destroyed, the volcano sending over 4 cubic miles of rock ash into the air. Eruptions like Tambora and

Krakatau distribute dust and debris throughout the planet and to such an altitude that it takes years for it to fall back to earth. Similar to the effects of Tambora, the eruption of Krakatau reduced average temperatures around the world for the next three years. It is estimated that the sun's intensity was decreased by 10 percent in Europe, and the appearance of the sun, moon, and sky was altered. Among other things, the material in the atmosphere was responsible for fantastic sunsets; some believe one is included in the background of Norwegian Expressionist artist Edvard Munch's painting "The Scream".

But if these examples fail to impress, one need only consider supervolcanoes, a less-discussed type of volcano, but one whose explosive power is so large they may be a threat to our very existence. The magma chamber for one supervolcano lies beneath Yellowstone National Park and is responsible for the active geysers and steam and hydrothermal vents one finds there. The last time this supervolcano exploded was 630,000 years ago, which is of some comfort. Considerably less comforting is what it did when it did erupt, which was to cover virtually all of the United States west of the Mississippi in a layer of ash. It left behind a caldera over 40 miles in diameter (it currently comprises virtually all of the park). The massive explosion is estimated to have been a thousand times more powerful than that of Mount St. Helens. Should it erupt today, it is difficult to imagine what life would be like in North America, or to what degree life would exist at all. And, we should be nervous. This supervolcano hasn't erupted for 630,000 years, but erupts roughly every 600,000 years. As Bill Bryson aptly writes in his book *A Short History of Nearly Everything*, "Yellowstone, it appears, is due."

But what does all of this have to do with lakes? How does the explosive force of volcanoes manifest itself in lake formation? The answer is simply that anything capable of creating a depression or hole in the ground can make a lake, and volcanoes are adept at nothing if not creating holes in the ground.

Nor is it just the holes at the peaks of volcanoes, the craters and calderas, that result in lakes. While impressive in size, these are actually small in number. Volcanoes generate lakes in large numbers via several other mechanisms including, for example, the formation of maars and pit craters, located all along the slope of a volcano and even in the flat land some distance from a volcano. The lava flows emanating from a volcano can also create lakes. River valleys are dammed, creating sometimes enormous lakes. Lakes can also form in the lava flows themselves, resulting, as we shall see, in numerous strange depressions that pockmark those solidified flows. Volcanoes create depressions and holes all over the planet, and wherever there is sufficient rain or groundwater or runoff, these holes will become lakes.

Volcanoes are not evenly distributed. In the continental United States volcanic lakes are found primarily in the west, particularly in the northwest where Oregon's Crater Lake and Hole-in-the-Ground Lake are just two examples. But, as we saw in the last chapter, lakes in North America, particularly in the northern part of the continent, tend to be of glacial, not volcanic origin. In other parts of the world, however, volcanoes dominate the limnological landscape. For example, in Japan the study of limnology itself is focused on volcanic lakes. Most of what is known about tropical limnology is due to the study of volcanic lakes in the volcanically active areas of Java, Bali, Sumatra, and Central America. Volcanic lakes are common throughout New Zealand, Iceland, the Auvergne district in France, and in the Eifel district of Germany. If we focus on volcanoes themselves, about three-quarters of all the active volcanoes on Earth can be found in what is referred to as the Circle of Fire, roughly coinciding with the edges of the Pacific Ocean, running along the Aleutian Islands, down the Kamchatka Peninsula, through Japan, the Philippines, Indonesia and New Zealand and then across the Pacific to the western edge of South America and on up through central America to the West coast of the United States and Canada. But this is only where recent volcanic activity exists. It is almost certainly true that there is no part of the planet that has not been subject to volcanism at some point in the geological record. So, while volcanic lakes are much smaller in number than those caused by glaciers, volcanic lakes can be found over a much wider geographical region.

The study of volcanic lakes lies at the intersection of limnology and volcanology and the resulting terminology and classifications is not entirely consistent between these fields (or even within them). Here, we classify volcanic lakes into two groups. The first consists of those lakes that form in the depressions on the volcano itself, either right at its peak or along its slope. The second group is lakes that evolve from the basins created by the flow of lava (or sometimes mud, and sometimes both) once it has solidified.

The first category includes lakes that are probably foremost in the minds of the popular imagination—a body of water occupying the crater located at the very peak of a volcano. The depressions formed at these peaks are sometimes referred to as craters and sometimes as calderas, a distinction that is far from agreed-upon. Most authors agree calderas are bigger than craters, with a dividing line at about a diameter of one mile. Many researchers typically refer to craters as depressions formed by the volcanic activity proper, that is, the buildup of ash, lava, rock, and other ejecta at the edge of the vent where the explosion emanates. This ejecta builds up to form a bowl-shaped depression that can turn into a lake once the volcanic eruption has ceased (this, as we will see, is never a guarantee). For explosions on the smaller end of the scale, such craters may be the end of the story. The volcano may erupt multiple times at the same location, or at nearby locations, creating a cluster of craters. But if these eruptions are relatively small, the resulting craters will be the defining depression created by the volcano. Once the eruption has ceased, lakes can form in all of these craters.

For larger volcanic eruptions, calderas can form. All volcanic explosions result from the existence of pressurized gas located beneath a vent. This gas may emanate from the magma where it was dissolved and subsequently came out of solution to form a gas. The gas can also be superheated steam, created when the magma encounters water ("phreatic" explosions, described in greater detail below). But in either case, when the gas pressure becomes large enough it explodes through a vent, often pushing out an enormous amount of lava in the process. This lava originates in a magma chamber at some depth beneath the vent and once the volcanic explosion is complete (which can take from minutes to days), the magma chamber is significantly changed. Its pressure has decreased, and, for large explosions, the volume of magma has decreased significantly, resulting in a large, unoccupied volume beneath the overlaying crust.

This process often results in the downward collapse of the crust, resulting in a massive depression called a caldera. Because the magma chamber can be much larger than the cone formed above it, the resulting depression, the caldera, may be far larger than the cone that formed during the actual eruption. Crater Lake in Oregon is an example of such a caldera. Crater Lake is huge (and, really, it should be called Caldera Lake), having a diameter of about six miles and a depth of almost 2000 feet at its deepest point; the distance from the rim of the caldera to the bottom of the lake is about 3900 feet. It is the second deepest lake in North America and one of the ten deepest in the world. Subsequent and less-violent activity has resulted in smaller craters on the floor of Crater Lake, one of which created a small island within the lake, Wizard Island, a visual illustration of the range in size of the structures that volcanoes can create.

Not all volcanic lakes are found at a volcano's peak. They can also form in depressions on the slopes of a volcano or on the flatter regions nearby. One example of such craters and one that often results in lakes are craters formed by phreatic explosions. These occur when magma rises from below and encounters groundwater. The groundwater flashes into superheated steam, which bursts up through the crust, creating a hole in the ground that may be away from the volcanic peak. Because the cause of such explosions is the groundwater already existing above the magma chamber, the resulting crater is typically filled with water, forming a lake fed by the water table.

Such lakes are called maars and are found throughout the world. They are especially prevalent in the Eifel region of Germany, where they have been well-studied. Maars are typically circular in shape, less than a half a mile in diameter, and less than 500 feet deep. In some cases, the superheated steam released during the explosion results in little lava release and so the debris cone around the lake is minimal, consisting only of the fragmented crust that was ejected during the explosion. This can make the lake difficult to identify, since it lacks the characteristic volcanic rock that surrounds crater lakes. Indeed, simply identifying the debris cone itself can be tricky, resulting in a lake that seems somewhat out of place, a hole in the ground with no apparent origin. Indeed Hole-in-the-Ground Lake, referred to earlier, is such a maar, a bit larger than typical, having a diameter of about one mile.

A lack of a debris cone can be even more pronounced for pit craters, craters that can range in diameter from tens of feet to about a mile and their depth from a few tens of feet to more than a thousand feet. Pit craters are not formed by explosions of any type, but rather by the sinking in of a part of the volcanic surface. The details are not entirely understood, but it is thought that pit craters form when underlying magma recedes downward, perhaps through release at a lower side vent or an opening on the slope of a volcano, thus draining magma from above, the surface collapsing when the magma leaves. It is also possible that pit craters are formed by vertical cracks formed as a volcano settles and that cause a collapse at the surface, or when a magma conduit eats away at the rock above, melting it so that there is only a thin layer of solid crust at the top, which subsequently collapses, forming a crater. Excellent examples of pit craters can be found along the Chain of Craters Road in Hawaii's Volcanoes National Park. Though a common feature in the study of volcanoes, pit craters are absent in limnology textbooks.

It seems likely that many of these have filled with water to form lakes. Indeed, so long as the floor of the pit is below the water table, lake formation seems almost guaranteed. It is possible that since these lakes also lack a debris cone of any kind, they are simply classified as maars in the limnological literature.

The second group of volcanic lakes are those formed some distance from the volcano itself and as a consequence of a lava flow emanating from the volcano. We should note that while an erupting volcano is a pretty incredible thing with its fountain of glowing molten rock that can rise thousands of feet into the sky, lava flows are pretty amazing, too. Though they may not be able to compete with the drama of the volcano they emanated from, lava flows make up for what they lack in drama with their sheer scale, for the lateral size of these flows can be simply enormous. The geological record contains many examples of lava flows that extend for dozens of miles and some for over 100 miles in distance and having thicknesses of over 100 feet. In terms of volume, the material comprising a lava flow can be mind-boggling. Some volcanoes in the Hawaiian Islands are estimated to have produced more than 16 billion cubic feet of lava. In Iceland, the lava flow that was emitted just from the Laki fissure in 1783 alone is estimated to have been 2.8 cubic miles in volume. Indeed, Iceland is so dominated by volcanoes that its bedrock is 90 percent igneous (solidified magma)—the entire island nation is essentially an enormous patchwork quilt of lava flows.

Perhaps the areal extent of lava flows on the planet should be no surprise. The beginning of the rock cycle starts with igneous rock (rock formed from magma), from which all other rock, sedimentary and metamorphic, are subsequently formed. Pick up a rock, and you can safely claim that its constituents were all once some kind of lava. But still, knowing that a vast swath of the planet's surface was created by flowing rivers of liquid rock can be hard to wrap our minds around. Geological maps of the northwestern portion of the United States (as well as many other parts of the world) reveal vast areas covered with basaltic lava flows. Indeed, it is estimated that the Roza lava flow in eastern Washington State covers 20,000 square miles and comprises 600 cubic miles of rock. It is difficult to imagine a single cubic mile of hot, orange, molten rock, let alone 600 of them.

Given the prodigious amount of material that flows from a volcano, it should be no surprise that these flows can dam up river valleys, thereby forming lakes in a way that is much faster than, but not dissimilar from, the lakes resulting from the buildup of glacial moraine at the end of a valley described in the previous chapter. One example of such a lake is Lake Kivu, which straddles the border of Rwanda and the Democratic Republic of the Congo. This is a particularly interesting lake because what is now Lake Kivu was once a tributary of the Nile River. Lava flows during the Pleistocene dammed up that river, causing the lake to form. Once filled, Kivu no longer drained into the Nile and now drains south toward Lake Tanganyika (itself the second deepest lake in the world at 4826 feet) and ultimately into the Congo River basin, entering the ocean on the far western side of the continent rather than the north as it originally did. Lakes of this form can be found all over the world and include Lac d'Aydat in France, Lake Bunyonyi in Uganda, and Lakes Pankeko and Penkeko in Japan, both of which were actually dammed up by mudslides formed when an eruption occurred in the side of Bandaisan in 1888.

Perhaps more interesting than the lakes formed when lava dams a valley are the lake basins formed *within* lava flows themselves. Surfaces and edges of lava flows can cool and partially or completely solidify, even as molten lava continues to flow beneath. Thus, when the eruption generating the lava flow ends, some parts of the lava flow may have solidified surfaces, with liquid lava still flowing inside, creating what is essentially a flattish rock tube within which liquid flows. The hotter lava located on the upper slope may drain through these tubes, emptying and spreading out at a lower location and solidifying. Once lava from the upper slopes has completely drained, the inside of these tubes empty, leaving large flat hollow structures. These vacant tunnels may be stable enough to maintain their shape once completely cooled, resulting in tunnels or lava tubes that are sometimes large enough to be explored on foot. A hikeable mile-long lava tube can be found at Lava River Cave in Coconino National Forest about ten miles northwest of Flagstaff, Arizona. In other cases, the draining of lava beneath the partially hardened surface of the lava flow will slump as the relatively soft walls and ceiling of these tunnels collapse. These depressions can become lakes.

Lava flows can be complex. Their braided streams bifurcate and rejoin, changing their geometry as the volumetric flowrate of the lava source changes and as the underlying topography over which they flow changes. So, it is not surprising that the lakes that form from these lava flows have complicated shapes as well. A wonderful example of such a lake system is Lake Myvatn in northern Iceland. This highly irregular lake has numerous islands within it that are also highly irregular in shape. Both the islands and the shore of Myvatn are pockmarked with circular depressions, themselves formed by the slumping process described above. Many of these depressions are filled with water, creating small lakes on the islands within the lake proper, as well as along its shores. It is a complex and magical-looking place, as is characteristic of lakes formed in lava flows.

• • •

As we have seen, the process of making a volcanic lake can be a violent one. But can the lakes formed by volcanoes themselves exhibit violence? This is certainly the case when volcanic lakes form in the craters or calderas of active volcanoes. If the volcano becomes active again, all of the water collected in the crater can be suddenly ejected. More often the water in such a lake will drain at a high rate of speed out of one or more cracks that develop in the side of the cone. This results in a hot mud flow called a "lahar" that can be extremely dangerous. There is one crater lake of especial note in this regard, the crater lake in the caldera of the Kelut volcano in Java, Indonesia. Water accumulated in this crater lake has been discharged in 1771, 1811, 1826, 1835, 1848, 1851, 1859, 1864, 1901, and 1919, the 1919 lahar implicated in 5110 deaths. These lahars occurred so frequently that engineers were brought in to design and build a pair of mitigating tunnels, the Ampera Tunnels. These successfully keep the lake drained to a low level, thereby minimizing the death and destruction caused by this particularly lethal volcano.

But what if the volcano is dormant? When viewing a placid lake that occupies the crater of a long-dead volcano, such as Crater Lake in Oregon, it is perhaps tempting to see this as the denouement of this story of volcanic destruction. Such lakes appear calm enough, nothing like the glaring red lava lakes that once occupied these water-filled depressions. And indeed, catastrophes related to lakes in dormant volcanoes pale in comparison to a full-on volcanic eruption. But still, as if inheriting the ways of the magmatic furies that formed their basins, some volcanic lakes seem to have a penchant for death—a tendency to destroy, even without any volcanic activity. And there is no better or stranger example of this than that of Lake Nyos, a volcanic lake located in the West African nation of Cameroon.

• • •

On the night of August 21, 1986, Lake Nyos erupted. During that night, the lake emitted not lava, not ash, not hot mud, but instead a massive cloud of cool carbon dioxide gas that silently raced down the slope, killing almost everything below. About a quarter of a cubic mile of carbon dioxide was released from Lake Nyos that night, traveling downhill at close to 45 miles an hour. In the nearby villages 1746 people died, most as they slept. In the town of Nyos itself, virtually every soul died.

Unlike many volcanic disasters, the Nyos event did not occur a thousand or even a hundred years ago. Occurring as it did in 1986, scientists were able to travel to the site within days. Accordingly, we have a detailed picture of what happened at Lake Nyos—one that is both terrifying and strange.

For the few survivors of the disaster, the situation they woke to must have extended beyond terror and into the horribly surreal. Some of the survivors did not wake for two days, and when they did, everyone around them had been killed—their families were dead and their neighbors were dead. Stumbling out of their houses, they could be forgiven for thinking that some otherworldly force had descended upon them and that the entire world had come to an end. Every living thing had died. Their chickens lay dead in the streets. Their livestock lay dead in the fields. The corpses of birds lay scattered randomly about. Even the insects were dead; rescue workers who arrived later noted the silence, the absence of insectile cacophony so common to equatorial Africa. It was not until days later that insect life reappeared, arriving at about the same time as the vultures that came from adjacent areas to feast on the bodies.

For some time the cause of the disaster was unclear. Other than the bodies, everything was normal. The sun was shining. The fields were green. Buildings were not knocked down. Nothing was burned. Initially some suspected a virulent epidemic that left only the few with natural immunity to live. But none of the outsiders and government officials who trickled into the villages became ill. It quickly became clear that something else was at play, and that something else turned out to be carbon dioxide.

The key to understanding the Lake Nyos explosion is to understand how carbon dioxide dissolves in water. All gasses have a certain solubility in water, a limit beyond which no more of the gas can be added. For carbon dioxide that limit is about one liter of the gas per liter of water at atmospheric temperature and pressure. Further attempts to add more gas to the water, for example by bubbling gas into the water, will have no effect. The mass of gas in each of the bubbles will stay the same as they rise and ultimately exit the surface.

The solubility of a gas in water can be increased by increasing the pressure and/or decreasing the temperature. And, high pressure and low temperature is exactly what one finds at the bottom of a deep lake. Like many crater lakes, Nyos is quite deep. At 682 feet, its bottom lies over two football fields in length below the surface. At this depth, the pressure is intense: 20 times larger than that at the surface, a pressure where the solubility of carbon dioxide is 20 times larger than at the surface. Since water can hold a liter of carbon dioxide per liter of liquid at atmospheric pressure, at the bottom of Nyos a single liter of water can hold an incredible 20 liters of carbon dioxide. This is rather a lot of carbon dioxide, and since virtually no light penetrates the nearly 700 feet to Nyos' bottom, it is also quite cold, likely dissolving yet more carbon dioxide.

But being *able* to hold a lot of carbon dioxide gas doesn't mean that a lake *will* hold that much gas. There must be a source for that carbon dioxide, and for most lakes that source is the air at the lake surface. So, for most deep lakes, the potential to dissolve carbon dioxide at depth is high, but the actual amount of dissolved carbon dioxide will be quite low since there is no effective way for the surface carbon dioxide to diffuse down to the lake bottom. But in the case of Lake Nyos, the carbon dioxide did not come from the surface. Though long dormant, some volcanic activity does exist deep beneath the floor of Lake Nyos, resulting in the formation of carbon dioxide. The gas seeps up through cracks and fissures in the rock, ultimately bubbling up through the lake floor, introducing it precisely where its solubility is highest.

Imagine these carbon dioxide bubbles, rising from the bed of Lake Nyos. As they rise, they encounter water that is very cold and at a very high pressure—water starved for carbon dioxide. And so, if you were at the bottom of Lake Nyos and could somehow see in the murky depths, what you might observe emerging from some crack in the lake floor would be a plume of bubbles rising upward and getting progressively smaller as they rose, the gas dissolving into the water until the bubbles simply disappeared. Imagine being on the floor of Lake Nyos, surrounded by numerous columns of bubbles rising upward and gradually disappearing, much the way the steam above your coffee cup disappears as it rises through the air above. What would be especially interesting about watching these bubbles disappear would be the dreadful knowledge that as each bubble dissolved, the lake would be getting closer and closer to disaster.

The process of dissolution of carbon dioxide into deep, cold water can go on for a very long time, perhaps a hundred years, a time when nothing untoward may happen. But bit by bit, the lake will become unstable. Year after year, decade after decade, the continuous dissolution increases the potential for disaster. Let's once again imagine we are at the bottom of Lake Nyos, this time with a bottle. If we fill that bottle with water at the bottom, seal it, bring it to the surface and then open it, the water will violently effervesce, foaming explosively out of the bottle. This is because water that is saturated at the pressures found at the bottom of Nyos becomes supersaturated at the far lower pressure of the lake surface. The carbon dioxide in the water at the lake bottom *wants* to be in gas form when suddenly exposed to the lower pressure at the lake surface, causing rapid degassing of the water in the same way that a bottle of soda may effervesce when opened. Now, what if instead of a person going to the bottom of the lake and bringing the bottom water to the surface, something else destabilized Lake Nyos, causing some of its deep water to move upward into the warmer, lower-pressure layers of the lake? Should this happen, the upwelling water would release its dissolved carbon dioxide in the form of bubbles. This sounds harmless, but in fact it would be catastrophic.

Nobody is sure quite what caused the bottom water to move upward in Lake Nyos. Some have hypothesized that small seismic tremors were the cause, or perhaps a sloshing of the water in the lake due to a wind of just the right speed. Others have posited that steam explosions deep beneath the floor of the lake (a form of phreatic explosion, described earlier) were the cause, or perhaps some kind of underwater landslide. Whatever the cause, some of the cold water from the depths of Lake Nyos did move upward on the night of August 21, 1986, and once this initial perturbation occurred, disaster was only seconds away.

As the cold lake water moved upward, its pressure dropped, and its temperature increased as it encountered the warmer water residing above it. This caused some carbon dioxide to come out of solution and form bubbles. By itself, this wouldn't have resulted in a disaster. However, the rising bubbles entrained behind them some of the liquid beneath, which in turn became supersaturated as it rose, releasing more bubbles. These bubbles entrained yet more deep water generating yet more bubbles. Thus a feedback cycle formed, each bubble resulting in the formation of other bubbles, each in turn creating more. Such processes go by a variety of names: self-reinforcing processes, positive feedback cycles, exponential growth, and others. Another term for this type of process is explosion. The cold, saturated water that had rested comfortably at the bottom of Lake Nyos for untold years rose to the surface at an explosive rate, causing massive quantities of carbon dioxide to come out of solution all at once. Though deadly, the gas release was relatively quiet. Those who survived reported little more than a rumbling, the sound of something like a distant explosion or rockslide. Most probably heard nothing at all.

Compared to the death tolls that are associated with volcanic explosions, those found at Nyos may seem small. But considering the absence of lava and fire and the darkening of the sun by ash—the fact that these deaths were due to nothing more than an invisible gas, the death toll was staggering. According to the United Nations, 1746 people died. At least 300 people ended up in the hospital, 3000 people were displaced, and 3500 head of cattle were killed. In the town of Nyos itself, there were only four survivors. To give a sense of the volume of gas released, some of the dead were found in villages as far as 12 miles from the lake; dead cattle were found at elevations as high as 330 feet above the crater rim.

That anybody survived at all is a miracle. Perhaps it was the oddities of wind and air flow that enabled some to receive enough oxygen to survive. Perhaps others were located in dwellings with minimal ventilation so that the carbon dioxide didn't come inside. Perhaps the survivors were in a high location in their dwelling where, like a person trapped in a sinking automobile, there was a bubble of air that allowed them to survive until the carbon dioxide dissipated.

Lake Nyos is now actively studied, and a degassing strategy has been implemented, aiming to prevent a repeat disaster. However, there are other crater lakes in Cameroon. It is estimated that there are 44 of these in Cameroon's Northwest Province alone, where carbon dioxide may be building in the depths. A similar explosion occurred at Lake Monoun, also in Cameroon, in 1984, where 37 people died. It is unclear if there are more carbon-dioxide lake explosions in Cameroon's future or whether such conditions may exist in crater lakes in other parts of the world.

But as bad as the Nyos explosion was, it should be noted that carbon dioxide, while lethal, is not combustible. The same cannot be said for Lake Kivu, the lava-dammed lake described earlier in this chapter. Kivu lies on the opposite side of Africa from Cameroon, bordering Rwanda and the Democratic Republic of the Congo. Like Nyos, gas is released from the floor of Lake Kivu, but in this case, that gas is not just carbon dioxide, but also methane, basically natural gas, like what is used in stoves and furnaces. Exactly what the consequences and risks this has for the people who live around Kivu is unclear. But one should probably take note of two facts: natural gas burns, and Lake Kivu is more than a thousand times larger than Lake Nyos. Due to Kivu's size and the high population density along the shores of the lake, some have stated that a Nyos-like event at Kivu would result in one of the largest natural disasters in human history. Volcanoes, it seems, present a threat to humans long after their eruptions have echoed into history, continuing to threaten lives along the cool shores of the lakes they create.

3. DAMS

Natural and Manmade

You would think that by now scientists would know how many lakes there are in the world, but we don't. You would especially think that we know how many manmade lakes there are since these require the construction of a dam and presumably some sort of official record. But, with the exception of relatively large lakes, we don't know how many artificial lakes there are either. Not only do we not know these numbers, but the estimates that we do have vary wildly.

As mentioned in Chapter 1, according to *The Lakes Handbook* (2004) the number of lakes on Earth larger than 2.47 acres in area is slightly more than eight million. But research published two years later in a paper where Iowa State University's J.A. Downing was the lead author, used new data sources and analysis methods and came up with 304 million. Partly this dramatically higher number was due to defining a "lake" as a water body larger than 0.247 acres. But even when considering lakes larger than 2.47 acres (the same as for *The Lakes Handbook*), Downing's work yields 26.6 million lakes, a figure over three times larger.

Subsequent studies have revealed yet more variations in the estimate. For example, a 2012 USGS study with Cory P. McDonald as the first author estimates that there are 64 million lakes in the world greater than 0.247 acres, significantly less than the estimate of Downing et al. for that size range. More recently, a 2016 study published in *Nature Communications* by a group at McGill University's Department of Geology estimates that there are 1.42 million lakes in the world. This last study is really not as small as it sounds given that it considered only lakes greater than 24.7 acres in area. For that size range, the McGill group's results actually fall close to those of Downing et al. Taken as a whole, these studies suggest that there are a lot more lakes on Earth than originally thought, but also significant disagreement in exactly how many actually exist.

That there is such disagreement in estimates of the number of lakes is probably not a surprise. It should be noted that nobody has ever tried to actually count the total number of lakes on Earth. In the studies described here, significant analytical work is done to extrapolate a relatively small data set in order to make a global prediction. For example, the McDonald et al. study used data on less than 1 percent of the estimated number of lakes on the planet in order to estimate the total number. And so it is not unlikely that future surveys will come up with different figures. Although surprising and perhaps somewhat concerning, our lack of knowledge here is also a bit heartening. In a world where satellites and drones seem to have mapped and identified just about everything, it is nice to know there are still some things about our planet that remain difficult to know and count—that there are still some surprises left out there.

Regardless of the degree of uncertainty surrounding the number of lakes in the world, we can say one thing with great confidence. We want more. As big as the number of lakes seems to be, it is nowhere near enough to satisfy our desire for these bodies of water. Humans want more lakes and, not surprisingly, humans love to build dams—to restrict the flow of a creek, stream, or river, and create a lake.

We humans are prolific lake-builders and we build lakes for myriad reasons. We build lakes to store water for irrigation. We build lakes to provide water for livestock. We build lakes to reduce flooding during periods of heavy rain. We build lakes to ensure that municipalities will have adequate water supplies during times of drought. We build lakes to run turbines that turn generators that provide electricity. We build lakes to provide cooling water for nuclear reactors and fossil-fuel-fired power plants. We build lakes for recreational use—to go fishing and swimming as well as to create lakefront property with its attendant docks enabling us to enjoy our boats and jet-skis. Sometimes we build lakes for ornamental reasons—to enjoy a reflecting pool or to have a small carp-filled pond in our backyard garden. We dig quarries and open pit mines, and sometimes when we are done with them, we let them fill with water, creating some very deep lakes. We build lakes for aquaculture—to raise quantities of catfish and salmon and tilapia to feed our ever-growing populations.

With dams we impound water that captures the tailings and other excreta of our mining industry. We build dams and locks to enable navigation of rivers and intracoastal channels that would otherwise be too shallow or too turbulent. We build dams to capture the runoff from irrigation, (which may have come from yet another lake). We build dams to create the lagoons of animal waste collected from pig farms. We collect and decompose human waste in sewage lagoons. We even build dams to capture water for energy storage, a system wherein a power plant pumps water uphill from an existing lake to an artificial basin during nighttime hours when electricity is cheap. This water is then released through turbines during the day, generating electricity that can be sold when the demand for electricity is higher (as is its price!).

Humans build a lot of lakes, and we have been doing this for an extremely long time. The oldest known dam is the Jawa dam, located in what is now northern Jordan. It is estimated to have been built at the beginning of the third millennium BC. Perhaps only 200 years later, Egypt's Sadd el-Kafara dam was built. Though both are quite old, evolving archeological research as well as the level of technologies exhibited at these sites suggest a much longer history of dam-building, perhaps stretching back as far as the seventh millennium BC.

The aforementioned ancient dams are no longer functioning, but many very old dams are still in use. According to the International

Commission on Large Dams (ICOLD), the oldest functioning dam is the Proserpina Dam in Extremadura, Spain, which was built by the Romans around AD 130 and is still in use today. And ICOLD lists at least 20 dams built before AD 1500 that are still in use.

For the most part, dams have served human beings very well, both those from antiquity and their more modern counterparts. However, as we will see later in this chapter, dams can be harmful in a variety of ways, not the least of which is that sometimes they collapse.

Humans continue to build dams, and at a brisk pace. It is estimated that from 1950 to 2000, the volume of water impounded behind manmade dams increased by a factor of ten, and continues to grow. This notwithstanding, estimates of artificial lakes are often quite small when compared to natural lakes. For example, the Downing et al. study mentioned earlier estimates 515,149 impoundments (water bodies created by a dam). And a 2008 study led by B. F. Chao of Taiwan's National Central University estimates 29,484 named reservoirs built since 1900. But it should be noted that these estimates are skewed toward larger lakes formed by large, engineered dams where good records have been kept. Far greater in number are the small impoundments used for irrigation, livestock watering, aquaculture, runoff capture, etc. for which few if any records are kept.

There exist few estimates of these smaller impoundments, and most studies are of a very restricted geographical area, making extrapolation to global numbers difficult. But the inventories of small manmade lakes that do exist suggest humans are very prodigious when it comes to building small dams. For example, the Virginia Cooperative Extension estimates over 50,000 ponds were built for farm use in Virginia alone, a number that, by itself, exceeds the value of 29,484 large impoundments for the entire world obtained from the Chao et al. reference cited above. Even larger numbers can be found for small-pond estimates in other states. Mississippi claims 160,000 small ponds, 200,000 are identified as fishing ponds in Tennessee, and the Iowa Department of Natural

Resources claims 110,000. Clearly, in terms of total numbers, small impoundments dominate.

Although an accurate global inventory of these small artificial lakes is probably not going to be available anytime soon, obtaining such estimates should probably be made a priority. One of the results obtained in the work of Downing et al. is that small lakes are far more important than once thought. Important ecosystem processes occur at the air/water interface of lakes such as the transfer of oxygen, carbon dioxide, and methane, as well as the sequestration of carbon. This makes important the total surface area of a lake or group of lakes. Downing et al. estimated that the total surface area of lakes is about 1.6 million square miles, which is 2.8 percent of the non-oceanic land area of the planet. This is a pretty big percentage considering that what we mean by non-oceanic land area is basically the same as dry land. In other words, 2.8 percent of the land mass above sea level is actually covered by lake water. Especially important is that more than 25 percent of this lake surface area exists in lakes smaller than 24.7 acres, and more than half the lake surface area exists in lakes smaller than 2470 acres. This suggests that the large lakes of the world (such as Lake Superior and Lake Baikal) do not dominate freshwater ecosystem processes as was previously thought. This revelation was somewhat of a surprise and suggests that greater attention should be paid to lakes, even (or especially) very small lakes, in global-climate models that had previously overlooked them. Simply knowing how many small lakes there are, and how many we are making, is clearly important.

What is also clear is that the number of dams continues to grow. This is certainly the case for small ponds used in agriculture. In the case of larger dams constructed for hydropower, though opportunities for new hydropower plants in North America and Western Europe are small since much of the opportunity for hydropower has been well-exploited, the same is not true in developing countries. For example, only 15 percent of the available hydropower has been exploited in India, and only 8 percent in Africa. When it comes to the construction of dams, large and small, humans have only just begun.

One might not be too surprised at the feverish pace of lake building in arid regions, but the growth in small impoundments is in no way restricted to such water-sparse locales. As noted earlier, even in the state of Mississippi, blessed with both prodigious rainfall and access to one of the largest rivers on the planet, small impoundments abound. In South Carolina where I live, which is also a relatively wet part of the United States, the average rainfall in the northwestern part of the state is 52.7 inches per year. This notwithstanding, even the briefest of drives in the country reveals a myriad of artificial lakes, ponds, and impoundments. The slightest trickle of a stream seems to tempt the average landowner to create a pond of some sort. Virtually every farm you see will have some form of impoundment, even if only for irrigation or the watering of livestock.

It is easy to see the temptation to make these impoundments, given their utility. And since construction of the required earthen dam necessitates little more than the services of a bulldozer operator, the making of such ponds is eminently doable. And this tendency to build lakes is by no means specific to South Carolina but can be seen throughout the globe.

The reason for this can be summed up by a saying you will hear if you spend any length of time with people who work in the area of water resources: there is plenty of freshwater available, it just isn't available in the right place. This saying is often invoked in the United States to bemoan the growth of populations in the desert Southwest while populations diminish in the Great Lakes region despite access to the thousands of cubic miles of water present in those lakes. But the saying applies on a very local scale as well. Water present in a river or lake even a few miles from where a landowner needs water will cost money to transport. Or, the right of access to that water is limited or nonexistent. But when a small stream runs through that landowner's property (or even just a small gulley that channels runoff during rains), the choice becomes obvious. This is the choice made by many, as is evidenced by the millions of small artificial impoundments existing today.

The consequences of all of these dams, both big and small, are many. Perhaps the most striking is the fact that some rivers no longer make it to the ocean. Rivers that once rushed to the sea as an awe-inspiring torrent of water now reach their end as little more than a meek trickle. This is due to the artificial impounding and withdrawal of enormous quantities of water all along the course of a river, water that is diverted for irrigation, municipal consumption, recreation, and other uses. In these situations, the demand for water matches or exceeds inflow into the basin, resulting in a flow at the river mouth slowing to almost nothing. The Colorado River is perhaps the best-known example. So oversubscribed is Colorado River water that when it meets the ocean at the Gulf of California in Mexico, it is almost nothing at all. A look at the mouth of the Colorado on the satellite view of Google maps reveals little more than a muddy channel. Instead of reaching the ocean the water collected in the Colorado River basin feeds the thirsty cities of Las Vegas, Phoenix, Tucson, Los Angeles, Denver, and others. This would not be possible without enormous dams impounding huge amounts of water. These include the well-known Hoover Dam and the Glen Canyon Dam. And it isn't just the mighty Colorado that has been reduced to a dribble. Powerful rivers that do not or at times do not reach the oceans or seas into which they traditionally flowed include the Indus and the Amu Darya in Asia, the Yellow River in China, the Murray River in Australia, and the Rio Grande.

As one might expect, all of these dams have an impact on the environment. Fish species that require a fast-moving water habitat may not survive in an impounded lake. The marshes and wetlands in flood plains are eliminated when dams are built. Indeed, a common reason for building many dams in the first place is for flood control. Once in place, a dam results in a slow and predictable rise in lake level even under very large rain inundations. Water can be released downstream in a controllable fashion that would not be the case otherwise. While this saves lives and protects property, the trade-off is that it eliminates the flood-plain habitats that many species rely upon.

Ironically, while some dams are created to husband water for future use such as irrigation, by taking the thin outline of a river and expanding it into the enormous area of a lake, evaporative loss of water increases. Of course in a river system without any impoundments, water, once it has flowed through the system and reaches the sea, cannot be used for irrigation, drinking water, or other uses. So, the increased evaporative loss caused by building artificial lakes is a loss of water that would have been lost anyway, and hence such impoundments still make sense from a human perspective.

But, from the perspective of the environment and the global water cycle, the increased amount of water impounded on continents by dams has a real effect. The work of Chao et al., cited earlier, showed that large artificial dams have impounded 2590 cubic miles of water and in so doing have actually had an impact on sea level rise, reducing it by 1.2 inches during the last half-century and at an average rate of -.02 inches per year during that time. Of course sea levels have actually risen during this period of time. The decrease in sea level caused by lake-building is not as large as the increases caused by the melting of ice from glaciers, thermal expansion of the ocean volume, and other factors, and so the net result is a rise in sea level. Nevertheless, it is still stunning to note that by keeping water impounded via the dams we build, the levels of the entire planet's oceans would be decreased by over an inch if other factors did not come into play.

The impact on the global climate caused by water impounded by dams is a story still playing out. The effect of the dams we've built and the ones we are building on sea level and the water cycle at large is something we are really only beginning to understand. However, we already understand quite well the impact that dams that we've built can have when they fail.

Water is very heavy. A gallon weighs a little over 8 pounds, and one cubic foot weighs a little over 62 pounds. When you start putting millions of cubic feet of water in one place, you have a lot on your hands. Normally this isn't a big deal, and a very deep lake can do little more than make your ears hurt if you dive too far into its depths. But the danger increases significantly if the lake exists behind a dam, if all of that water resides behind an earthen or concrete construct, a dam that could fail. If this dam is a small earthen embankment holding back a livestock pond covering a few acres of rural farmland, the consequences will be small. But, for big dams, the consequences of dam failure can be catastrophic.

The worst dam failure in the United States was the 1889 Johnstown flood in Pennsylvania, where over 2200 people died. The Johnstown flood was and is widely memorialized, having been the subject of numerous novels, plays, poems, and even a *Star Trek* episode. Happily, this disaster has remained the exception for dams in the United States where such large-scale failures are rare. However, there is still good reason to worry about dams in the United States since there are a lot of them, and some are in bad shape.

Moreover, the condition of the nation's dams has been worsening. The American Society of Civil Engineers (ASCE) keeps track of dams, classifying their hazard potential as low, significant, or high. According to their 2017 Dam Report Card, as of 2016, nearly 15,500 dams fell into the category of high-hazard potential, meaning their failure is anticipated to cause loss of life. This number was only 10,213 in 2005. The problem is only expected to grow as maintenance on dams gets progressively deferred, a significant concern given that at the time of the 2017 Dam Report Card, the average age of these dams was 56 years.

But as bad as it can be when a manmade dam fails, we have a fighting chance at reducing, if not eliminating the potential for catastrophe by maintaining these dams and by avoiding development in regions beneath dams. The same cannot be said for landslide dams, which can and do cause incredible disasters and for which very little can be done.

When a landslide blocks a river, a lake can form. These lakes tend to be very short-lived, though there are notable exceptions to that rule. One is Lake Sarez in Tajikistan, formed by a landslide in 1911 and stable ever since, peacefully holding back almost 4 cubic miles of water without incident. But the vast majority of landslide dams, sooner or later, are almost always disasters.

The progression of events is awful. First, there is a landslide, sometimes caused by excessive rains but more typically by an earthquake (and in especially bad situations by excessive rains followed by an earthquake). People are killed when the wall of soil, rock, and trees comes falling down a mountainside and then blocks a river. But the awfulness has only just begun. Upstream of the dam, a lake forms, quickly filling up the valley and displacing or killing the residents of the villages in the valley. As the lake fills, the stability of the dam comes into question. They are not, after all, built for the job, but they hold water back long enough for a temporary lake to fill. Often these dams don't fail early in the process when, say, the lake is 10 percent filled, or 50 percent filled, when the damage of failure would be far less. Instead, perversely, landslide dams often hold back all the water that they possibly can—and *then* fail, the water rushing over the top. This is a mechanism called *overtopping,* a process that quickly destroys the dam and sends a massive cascade of water downstream, destroying or carrying away everything in its path.

One of the worst landslide dam disasters on record was the Tangjiashan landslide dam failure. This disaster followed the Wenchuan earthquake in China, a magnitude-8.0 quake that struck on May 12, 2008, in a mountainous region in south-central China. It leveled buildings and caused immense destruction to eight provinces and cities, including Sichuan, Gansu, Shanxi, Chongqing, Yunnan, Shaanxi, Guizhou, and Hubei. It caused 69,200 deaths and destroyed over 5 million homes. The earthquake also created an incredible 200,000 landslides, which in turn created 257 landslide dam lakes.

The misery and fear created by this earthquake must have been overwhelming. As if the main earthquake itself wasn't enough, over 20,000 aftershocks followed, surely terrifying the survivors. The rainy season in this area begins in July, two months after the earthquake. And so, as those living in the area attempted to pick up the pieces of their lives and bury their dead, the rains began, and the landslide dam lakes began to fill. As the 257 lakes filled, the possibility of an uncontrolled outburst from these lakes presented a growing risk to the lives and property of more than 130 million people in the downstream region. The Tangjiashan lake was the largest and most dangerous lake formed during the Wenchuan earthquake and had the potential to impact over one million people in Mianyang, located downstream. The amount of water impounded behind the dam was enormous, with an estimated water storage capacity of 11.1 billion cubic feet; it had submerged the valley to a distance of nearly 15 miles upstream. The landslide dam itself was also large, measuring over 2000 feet in the cross-river direction and 2635 feet along the length of the river. The height of the dam ranged from 270 to just over 400 feet.

The response of the Chinese government to the Tangjiashan landslide dam was overwhelming, and for good reason. The potential for disaster posed by the Tangjiashan dam was clearly significant, and Chinese history is rife with disasters due to the failure of landslide dams, surely amplifying the perceived danger of the Tangjiashan landslide dam. The biggest landslide dam disaster ever recorded in China occurred on June 1, 1786, in the Kangding-Luding area of southwestern China where a 7.8-magnitude earthquake occurred causing a landslide that blocked the Dadu River. The resulting lake existed for 10 days, during which an estimated 1.8 billion cubic feet of water accumulated behind the dam. An aftershock of the original earthquake breached the dam on day 10, and the resulting torrent of water was enormous, estimated at over 1.3 million cubic feet per second at its peak, a value comparable to the Mississippi River at flood. Historic documents reveal that over 100,000 people were killed by the flood, making this one of the deadliest landslide-dam failures in history.

More recently, China has had other landslide-dam failures that were quite significant, though not quite on the scale of the 1786 event. On August 25, 1933, a magnitude-7.5 earthquake struck in Sichuan Province near the town of Diexi. Three landslide dams formed on the Min River. As river water accumulated, the upstream dams overtopped and, because of the relatively high height of the next dam downstream, these three lakes merged into one. The downstream dam held for 45 days, then it overtopped, was destroyed, and caused a massive flood that traveled an incredible 155 miles downstream, killing over 2500 people. More recently, on June 8, 1967, a huge landslide occurred, also in Sichuan Province at Tanggudong, blocking the Yalong River. The dam lasted for nine days. When the dam overtopped, the resulting flood had a peak flow rate of 1.9 million cubic feet per second. This time no lives were lost—the Chinese government, having learned the lessons of previous disasters, had evacuated the downstream population.

And so, presumably to ensure history would not repeat itself, the Chinese government brought to bear just about everything it could when confronted with the landslide dams formed during the 2008 Wenchuan earthquake. To prevent disaster, a landslide dam lake must be drained before the dam fails catastrophically. This is a relatively tricky procedure. One wants to controllably drain the lake by removing part of the dam. But, there is always the risk that as the water begins to flow, rapid erosion will make any small channel into a bigger one, a process that can grow exponentially. In any event, the process requires significant resources. To address the problem, the government sent over a thousand soldiers to the site, along with heavy excavation equipment. The Tangjiashan landslide dam was located in a relatively inaccessible region made more so by the destruction of roads via the landslide. Equipment and personnel were helicoptered in or transported upstream in rubber boats. As a precaution, 275,000 people were evacuated from the downstream area before the dam was destroyed.

On May 21, over a thousand pounds of explosives were detonated to create a channel in the dam. A relatively small channel was made, and the lake began to drain, but slowly. Between June 10 and June 11, three underwater explosions were implemented at the drainage channel cutting through Tangjiashan dam. A final detonation was conducted on June 12. In all, 14 tons of explosives were used. A plot of the drainage channel outflow versus time reveals the dicey nature of draining a landslide dam lake, one that was far from perfect in the Tangjiashan case. Prior to the final detonation, the flow through the drainage channel was small, barely visible. The final explosion was set at 10:20 a.m., yet the hydrograph showed almost no flow at 11 a.m. and still almost no flow at noon. Then, shortly after 1 p.m., the flow rate skyrocketed, ramping up to a peak of 139,500 cubic feet per second in the course of just minutes. Then, by sundown, the flow reduced to almost nothing. The lake had been drained in one mighty torrent, essentially in the course of a single day. If it hadn't been for the herculean evacuation efforts, the number of dead would have been enormous.

But this is the challenge of removing these dams. If you go slow and try to remove water at a very controlled rate, you run the risk of the dam failing on its own and only after the lake surface has attained a relatively large height. If you remove too much of the dam with large amounts of explosives, you may simply mimic the catastrophic failure you were trying to avoid. The goal is to find some middle ground, always difficult to attain in inherently unstable systems, as landslide dams surely are. At the Tangjiashan dam, perhaps something close to this middle ground was achieved, although one hopes a better result can be attained in future versions of this situation. For what was achieved on June 12, 2008, might best be defined as a controlled flood.

4. THE CAROLINA BAYS

The formation mechanisms of the lakes described in the previous chapters are fairly well understood. To be sure, there are aspects of volcanoes and glaciers that are still a mystery, and exactly how lakes are formed by these volcanoes and glaciers still require further study in some cases. But, that a volcanic lake was indeed born of a volcano is clear, and that a kettle lake is really the result of glacial processes is similarly well-accepted. But for some lakes, the very mechanism of their creation is simply unknown. These lakes exist, but like a child orphaned on the doorstep of a stranger, the circumstances surrounding their birth remain a mystery. This lends such lakes an air of mystique, making them more interesting than they otherwise might be. And this is nowhere truer than for the intriguing elliptical lakes found in the coastal plain of the Carolinas—a series of lakes called the Carolina bays.

If you have a computer or smartphone nearby, open your favorite mapping app and enter "Elizabethtown, North Carolina." Select the satellite view and zoom out to a level where the width of your screen is about 20 miles in scale. Look to the northeast of Elizabethtown, and gradually something incredible will emerge from the landscape. One of the first things you will notice is the presence of several large, elliptically shaped lakes. Among these are Salters Lake, White Lake, Jones Lake, and Bay Tree Lake. The water-filled portion of these lakes may not be elliptical, but if you look carefully, you will see an elliptical boundary, the non-water area filled with vegetation, a sandy ridge often delineating parts of the boundary. Near these lakes are other elliptical regions, most green, some filled with wild vegetation, others filled with the cross-hatched patterns indicative of tilled fields. The more you look, the more of these elliptical regions you will see, until you realize that the entire area is covered with these oval features. The density of these patterns is so large that they overlap in some places, and in others one elliptical region may fall entirely inside another. The surprise and mystery of the landscape is enhanced when you realize that these ellipses are all pointing in the same direction, their long axes oriented to the northwest, as if pointing to something, as if intimating where they came from.

If you have conducted the above geographical task, you have discovered for yourself a type of lake called a Carolina bay. Two questions that most likely ran through your mind were: "What are these things?" and "How did they form?" When I first observed these lakes from the satellite view of Google Maps, I asked myself these same questions. But in addition to asking myself questions regarding the formation, orientation, and ecology of these mysterious lakes, which will be described in greater detail later in this chapter, I also found myself wondering how I'd never heard of them before.

I am always drawn to unexplained phenomena, and the Carolina bays fall in this category. But my sense of wonder grows exponentially when an unexplained phenomenon exists virtually in my backyard, something that has been right under my nose the whole time. Here were these strange lakes, perfectly oval, as if made by some extraterrestrial life form. Here were lakes whose characteristics seemed to reveal more facets of themselves the longer you studied them. And yet they were only a few hours' drive from my home. The Carolina bays were a secret hidden in plain sight, and a fascinating secret at that.

There are many compelling aspects of the Carolina bays. There is the controversy surrounding how they formed, the unique species living inside them, and the question of why they all seem to point in the same direction. But let's start with the simple facts, the things we know with greatest confidence.

Carolina bays are lakes or wetlands, not literally bays, taking the second part of their name from the bay trees that frequently grow in and around them. Researchers often refer to them as "depressions" or "wet, low-lying regions," perhaps avoiding the word "lake" because many of them contain water only seasonally, and because many were long ago drained for agricultural purposes and no longer contain water at all. Carolina bays are not found solely in the Carolinas, though it is true that they are found in especially large numbers in the Atlantic Coastal Plain near the South Carolina/North Carolina border, particularly in Cumberland County, Bladen County, and Robeson County in North Carolina. However, they have been observed from as far south as northern Florida to as far north as southern New Jersey.

Estimates of the total number of Carolina bays vary widely, from a few tens of thousands to as many as 500,000. These estimates have increased over time as methods for identifying and counting bays have improved, from the early days of aerial photography, to satellite imagery, and now to the use of light detection and ranging (LiDAR), which enables identification of bays even when they are filled with and surrounded by thick brush and trees. This will likely result in further increases in estimates of the total number of bays. LiDAR is also likely to improve measurements of the shape and orientation of bays, which might in turn help confirm or deny theories regarding mechanisms for their formation.

There are two broad categories of research on Carolina bays. The first concerns the mechanism of their formation and orientation. The second area of research focuses on the unique ecology of the Carolina bays, which is far better understood than formation mechanisms, although there is still much to learn here as well.

Clearly the Carolina bays provide an environment where life tends to flourish in showy profusion. Research on these wet depressions reveals that there is a greater diversity of species found inside of them than in similar habitats very close by. In particular, studies often show large numbers of amphibians and a diversity of amphibian species. This is an ecologically important fact considering the precipitous decline in amphibians occurring worldwide. Some bays contain water year-round, enabling the presence of fish. While this does not preclude amphibians, fish often prey upon these creatures and their eggs, limiting their numbers. Most Carolina bays exhibit seasonal variations in water level, becoming completely dry during the summer. This prevents the existence of fish, a situation that enables amphibians and other vulnerable freshwater creatures to flourish.

Many of the amphibian species are threatened, making the bays a kind of naturally occurring preserve. In a 1997 *Smithsonian Magazine* article by Kevin Krajick, herpetologist Whit Gibbons of the University of Georgia revealed he'd once counted a staggering 200,000 leopard frogs on the shoreline of a single Carolina bay. While it is likely that ongoing development of regions in and around Carolina bays has negatively affected their ecology, current research shows existing ones remain extremely well-endowed in amphibians—descriptions of the environment in a Carolina bay invariably reveal a kind of microclimate rich in wildlife, a lush place alive with all manner of birds, plants, and insects. Carnivorous plants, while sometimes found outside of these bays, are found in profusion inside including not just the magnificent Venus flytrap, but also the sundew, pitcher plant, and the bladderwort.

It is oft noted by researchers that each Carolina bay seems to be different from the next. Perhaps species within each bay have evolved somewhat separately due to the slightly different environment in each bay. It is almost as if the sandy rims outlining each of these wonderful lakes or wetlands are sufficient to maintain them as independent petri dishes, separate worlds where evolution occurs as though each one was isolated.

This variability is perhaps not surprising since, unlike other kinds of lakes, the Carolina bays typically do not have inlets or outlets. Usually there are no streams or rivers to supply or drain them; water gets in via either rain or some subterranean water source. The upshot is that they are much more isolated biologically than they would be if water from a stream or river connected one bay to another or to any other body of water. In fact, some bays are isolated enough to have actually evolved distinct species that exist exclusively in that one bay. For example, Lake Waccamaw, one of the largest Carolina bays, 3.5 miles wide by 5.2 miles in length, is home to six species of fish and mollusks found nowhere else in the world. This is especially exciting because research on the biology of the Carolina bays is just beginning. With the enormous number of bays, biologists have only scratched the surface in cataloging their flora and fauna.

• • •

I first learned of the Carolina bays from a novel. *Nowhere Else on Earth*, a work of historical fiction by Josephine Humphreys, takes place in Robeson County. It is the fictionalized account of the real-life Henry Berry Lowrie, a Lumbee Indian who fought a guerilla war in and around Robeson County against the Confederates and the North Carolina Home Guard during and after the Civil War. Prominent in the pages of Humphreys' book is a detailed description of the land:

> *The county has more farms and towns now, more ditches, more roads and bridges and railroad tracks. But back then everything not pinewoods or fields was swamps, fifty of them labeled on the map and more whose names were never known to mapmakers. Some were pocosins, shallow egg-shaped basins landlocked and still, scattered northwesterly as if a clutch of stars had been flung aslant in one careless toss from heaven, leaving bays that sometimes filled with rain and sometimes dried in the sun, growing gums and poplars and one tiny bright green plant found nowhere else on earth, the toothed and alluring Venus flytrap.*

Humphreys' quote is wonderful, not just for how it captures the beauty of Robeson County, but also because it captures so much of what is fascinating about the Carolina bays (referred to as pocosins, one of the many names used for Carolina bays). And while the bays were not formed by stars flung from heaven, for some time there was a theory of their formation that did not deviate much from this description and that caught the attention of the scientific community and the media in a way few scientific topics do.

It is important to keep an open mind regarding the origin of Carolina bays, as there is no lack of number or diversity in the various theories. Thomas E. Ross, professor emeritus at the University of North Carolina in Pembroke (located within Robeson County), wrote an annotated bibliography in 1987 listing over 350 scientific articles on the Carolina bays and documenting 18 theories on their genesis, published from as far back as 1848 to the present. These range from the mundane: "Bays are sinks over limestone solution areas streamlined by groundwater," to the absurd: "Black hole striking in Canada throwing ice onto coastal plain," to the whimsical: "Fish nests made by giant schools of fish waving their fins in unison." Most of the theories catalogued by Ross were penned fairly far back in time and were subsequently discarded. Virtually all of the modern theories fall into one of two categories. The first is some variation on the idea that the bays were formed as the direct or indirect result of a meteor impact. The second group of theories claims the bays were initially randomly shaped depressions that were transformed into oval shapes by strong prehistoric winds blowing in the same direction for thousands of years. These unidirectional winds caused a water flow that eroded the shorelines into oval shapes, the orientation of the oval determined by the wind's direction. Both theories have their flaws.

Modern meteor theories for Carolina bay formation do not suggest that the bays were the result of direct meteor impacts wherein each bay was the result of a separate meteor. Indeed, a direct impact by a meteor would be extremely unlikely to result in an oval shape. Just as the craters we see on the moon are circular, so too are craters formed by direct meteor impacts on Earth, regardless of the angle of attack of the impactor. There are dozens of meteor-impact locations on Earth that demonstrate the circular nature of the resulting craters, though many are difficult or impossible to see due to their age and to the geological processes that have occurred since their formation. Three well-preserved meteor craters, easily seen on Google Maps, are the Wolfe Creek Crater in Australia, the Barrington Crater in Arizona, and the Pingualuit Crater in northern Quebec. All are inarguably circular.

The reason for the lack of elliptical craters from meteor impacts is a bit counterintuitive. One imagines a meteor hitting Earth at an angle would cause an oval depression as it moves along the surface during the impact. But that scenario presumes an impact that is far less violent than meteor impacts actually are. As noted in an excellent 1999 article in *Scientific American* by Gregory A. Lyzenga, the kinetic energy of meteors is immense. These objects have velocities measured in miles per second. The enormous kinetic energy of the meteor is converted into heat during the fraction of a second that comprises the entire impact event. During this extremely brief interval, the enormous amount of heat released dramatically increases the temperature of the meteor, thereby vaporizing it and causing an explosion that radiates outward from the impact location. This explosion pushes the dirt and rocks outward, and this ejecta forms a circular crater. Because the explosion is so powerful, the crater geometry is not affected by the angle of impact. Indeed, as noted by Lyzenga, the only way a non-circular crater could form would be if the meteor traveled at such a shallow angle that it moved a significant horizontal distance during the brief period of time comprising the explosion. In such an extreme situation, the explosion would propagate outward from a continuously moving point during the lifetime of the explosion. At such a low grazing angle, the resulting crater could be oval in shape, or possibly rod-shaped. But such crater shapes would require the meteor trajectory to be no more than a degree or two off from horizontal. While such an event is possible, it would certainly be rare. Furthermore, for a direct impact to be responsible for the shape of the Carolina bays, hundreds of thousands of meteors would have to strike the planet, all at that same very shallow angle of impact and all traveling in the same direction. The astronomically low probability of this scenario has resulted in the rejection of any Carolina bay formation mechanism due to multiple, *direct* meteor impacts.

In order for a meteor to create the Carolina bays, an indirect mechanism would be required. In this scenario, there would be a primary meteor impact site very far from the Carolina bays, and the ejecta material flung outward from this primary meteor impact would subsequently fall back to earth, forming the Carolina bays via secondary impacts. The mass and velocity of these secondary impactors would be smaller than that of the primary meteor. Hence, these secondary impacts would have a lower kinetic energy, perhaps low enough to avoid an explosion. Such relatively low-energy impacts could then result in craters that would not necessarily be circular and whose shape would be determined in part by the impact angle of the ejecta pieces.

The above suggests (but certainly doesn't prove) that Carolina bays could have been formed from the ejecta of a meteor striking some distance away in North America. It is a tempting hypothesis—one that tempted R. B. Firestone, a scientist at California's Lawrence Berkeley National Laboratory. In 2007, Firestone, along with 25 co-authors, published a paper in the prestigious *Proceedings of the National Academy of Sciences*, positing that 12,900 years ago an extraterrestrial object hit or exploded over North America. In this paper and in later work by Firestone and his co-authors (sometimes without him), it is stated or implied that one of several consequences of this impact was the formation of the Carolina bays.

Firestone's theory was broad in its consequences. The theory focused on a time in our planet's history referred to as the Younger Dryas period, from 12,800 to 11,500 years ago. This period occurred after the glaciers of the last glaciation had reached their maximum size and had been receding for some 7000 years. The Younger Dryas was a period of sudden cooling that interrupted this warming trend and coincided with the extinction of North American megafauna (large mammalian species), including the saber-toothed tiger, the wooly mammoth, mastodons, and the giant short-faced bear. It is also at about this time that evidence of the existence of a group of Paleo-Indians referred to as the Clovis people disappeared from the fossil record. This evidence consists of arrowheads and spear points having specific geometrical characteristics. Explanations for the extinction of large mammals and the disappearance of evidence of Clovis culture have long been controversial, and so any new theory was bound to be somewhat contentious. However, existing theories were far more sober than Firestone's. For example, one group of theories suggested the megafauna were simply killed off by overhunting, and that the Clovis people did not die off, but rather their arrow points merely evolved into other styles; that they were not indicative of other peoples, but were simply a natural technical evolution of the Clovis people.

On the other hand, Firestone and his coauthors (referred to hereinafter simply as "Firestone") hypothesized that an extraterrestrial object (such as a meteor) exploded over the Laurentian ice sheet in what is now the Great Lakes region of North America, causing fierce wildfires that raged over the continent, and caused the extinction of the Clovis people and the megafauna. As partial support for this idea, Firestone cited the existence of a "black mat" in the fossil record at a number of archeological sites in North America (including Carolina bay locations). Below this black layer exists megafauna fossil evidence and evidence of Clovis culture. Above this layer, such evidence is not observed. He suggested that the black mat was caused by the very wildfires responsible for the extinction of megafauna and Clovis people.

Firestone's compelling theory, often referred to as the Younger Dryas Impact Hypothesis (YDIH), became something of a media juggernaut.

It spawned numerous popular science articles and was the subject of at least three documentaries, including ones produced by PBS's *Nova*, The History Channel, and the National Geographic Channel. The YDIH captured the public's imagination. Perhaps this was because of its similarity to another meteor impact, the one understood to have resulted in the extinction of dinosaurs, an idea originally hypothesized by the father and son team, physicist Luis Walter Alvarez and geologist Walter Alvarez along with other co-workers in 1980. However, while the Alvarez hypothesis was largely confirmed by myriad other research groups and is widely accepted today, research from groups not associated with Firestone's original team eventually contradicted virtually all of what Firestone claimed. The details of this go beyond the scope of this book, but they involve the existence of various "markers" that would confirm the existence of an impact or the existence of wildfires spanning North America, as Firestone claimed. These markers take the form of certain elements and minerals one would expect to find in meteors or to be formed by meteor impacts. Firestone found abundant evidence of such markers, while unaffiliated groups found little if any. There were numerous iterations on Firestone's theories where he and his colleagues responded to criticisms of the hypothesis and modified their theories to accommodate new evidence or the lack thereof. The back and forth between Firestone and groups that critiqued his work went on for about a decade, yielding journal articles that were colorful in ways that one rarely finds in the scientific literature. Titles from that time include: "Experts find no evidence for a mammoth-killer impact,"; "The Younger Dryas impact hypothesis: a cosmic catastrophe,"; and "The Younger Dryas impact hypothesis: A requiem."

The controversy was fueled by odd personal revelations concerning some of the authors. One, a man named Alan West, began his life as Allen Whitt and kept his birth name until he was convicted by the State of California. Evidently he had falsely claimed to be a state-licensed geologist, charging towns to perform water studies under false credentials. Moreover, prior to the YDIH theory, Firestone and West had written a book together, *The Cycle of Cosmic Catastrophes*. The book expounded on cosmic impacts and their effect on Earth and has been referred to as pseudoscience. But personal oddities of the authors aside, as noted by David Morrison, NASA senior scientist and Committee for Skeptical Inquiry Fellow in a 2010 article in the *Skeptical Inquirer*, "A good hypothesis naturally accretes confirmation and gets better with time, as did the Alvarez ... impact hypothesis. Firestone's work has not done so." And as of today, there still hasn't been confirmation of the YDIH by researchers outside of Firestone's group.

But what does all of this have to do with the Carolina bays? Carolina bays play a small role in Firestone's work. A careful reading of Firestone's first paper reveals that the authors primarily use early work by University of North Carolina geology professor W. F. Prouty, who had suggested in a 1952 article that Carolina bays were formed by the shock waves of incoming meteors, mostly for the purpose of justifying use of the Carolina bays as sites for searching for markers of the YDIH impact. In other words, Firestone and co-authors presumed the Carolina bays were the result of the meteor impact and simply chose to obtain soil samples there.

But notwithstanding the fact that Carolina bays were not at the center of the YDIH hypothesis, the bad press surrounding it seems to have tainted the whole notion that a meteor could have had any role in their formation. When I discussed with a geologist colleague of mine the notion that secondary ejecta from a large impact site might have caused the Carolina bays, he did little more than give me a sad smile, a mild shake of his head, and remark, "No. There just aren't any markers there." It seemed his comment wasn't directed at my suggestion, but rather at all of the negative publicity surrounding Firestone's work. It was all the same thing in his mind, and this seems to be a common

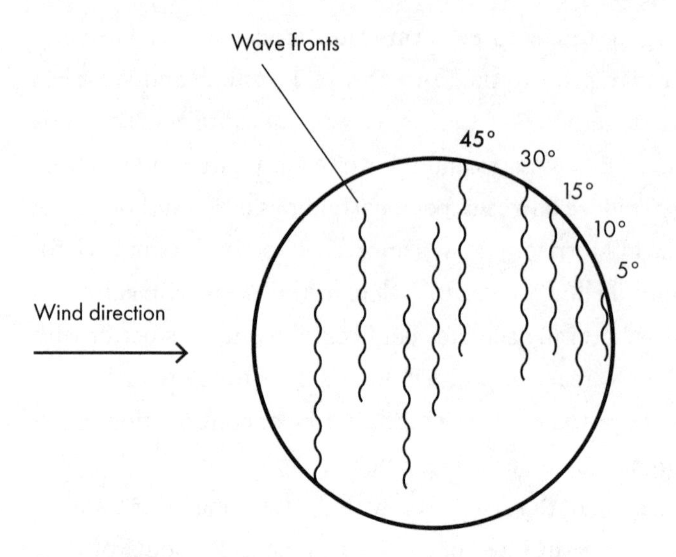

Figure 1. Diagram showing wave fronts and their angle of incidence on the downwind shore. Maximum shore erosion occurs when this angle is 45 degrees. The initial lake geometry is arbitrarily chosen to be circular. After Kaczorowski, 1977.

attitude when it comes to the Carolina bays in these post-Firestone years. But if the Carolina bays were not formed by a meteoric impact, then what is responsible for their creation?

The theory that seems to have garnered the greatest support in the geoscience community to date goes back to a technical report submitted by geologist Raymond T. Kaczorowski to the Department of Geology at the University of South Carolina in June 1977. Kaczorowski largely disappeared from the geology literature shortly after. However, the influence of his report is large; it is difficult to find a publication on Carolina bay formation that lacks a citation of Kaczorowski's work. Kaczorowski suggested that at the time of Carolina bay formation, what is now the southeastern United States was undergoing a long period of strong, unidirectional winds. Under these conditions, he postulated, a process occurs, similar to what is seen in oriented lakes. Oriented lakes are further discussed in the next chapter and are found in Northern Alaska as

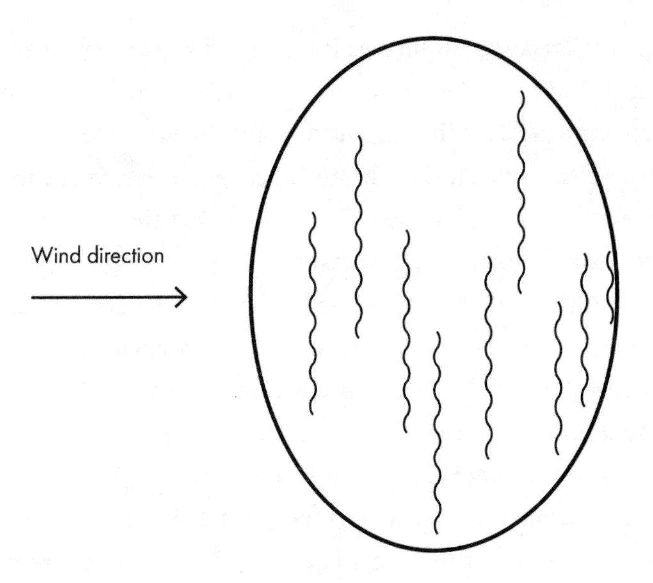

Figure 2. Resulting shape of a Carolina bay according to Kaczorowski's hypothesis.

well as in Texas and Chile. These lakes are understood to begin as arbitrarily shaped water-filled depressions, but are transformed into oval shapes by winds that are either unidirectional, or bi-directional (coming intermittently from opposing directions). In this lake-formation mechanism, wind forms waves that propagate across the water surface and impact the far shore, which they erode. A maximum in wave erosion occurs when the angle between the direction of wave propagation and the shoreline is 45 degrees. As shown in Figure 1, this angle is achieved at the portions of the shore close to 90 degrees from the upwind and downwind shore locations. Continued wave action extends the lake in this direction, creating a lake that, according to Kaczorowski, would be oval in shape, as shown in Figure 2.

Kaczorowski's contribution was to compare the Carolina bays to oriented lakes in other parts of the world and to suggest that the mechanism that formed oriented lakes is identical to what formed the

Carolina bays. Kaczorowski further investigated the feasibility of this mechanism by building a model, a sand table with a water-filled depression. By blowing air over the table using fans, he was able to test his hypothesis. At least one photo in his technical report shows the formation of an outline in his sand table that is, while not the perfect ellipse one sees in many Carolina bays, at least somewhat oblong.

While the theory of uniform winds has enjoyed the greatest degree of acceptance, it also has many drawbacks. One particularly challenging observation is the fact that it is not uncommon to see a bay existing inside of another bay, which would be difficult to explain using a wind-based formation mechanism. Such a pair of bays would require the outer bay to be formed first, since the inner bay would have been eroded away by the water flow if the outer bay formed second. This requires that once the outer bay was formed, the water level dropped suddenly and stayed at a much lower level for an extended period of time, at least enough time for the inner bay to form. For the water level to remain at such a constant low level for such a long period seems unlikely. Water levels in any kind of lake or pond typically display variations over yearly, seasonal, or even weekly time-scales. Since Carolina bays tend to lack inlets and outlets, they are likely to experience significant changes in water levels due to precipitation making it unlikely any single fill level would exist for long periods of time. In order for the water level in a bay to drop and stay low enough to permit formation of a small bay within a larger one, it would require centuries of stability in water level. This possibility seems difficult to imagine.

Yet more difficult to explain is the existence of overlapping bays, which is common in regions where Carolina bays exist in high concentration, something easily confirmed by the most casual look at satellite imagery of the Robeson County area. If the uniform-wind theory is correct, then two overlapping bays would have to form in sequence, the first bay being the one whose oval boundary is penetrated by the second bay, whose boundary is unbroken. But in this situation, one of these bays would have to have water in it, while the other did not. In other words, there would have to be a very long period of time when the first bay was filled with water, followed by another long period of time when the second bay was filled with water while the first was not. How this could happen is unclear. Both bays would receive the same amount of rainfall, which would presumably result in both having comparable water levels. If there was a long drought with little rainfall, then subterranean water might be a source for the bays. But it is difficult to imagine how subterranean water would be available to one bay but not the other, given that they overlap. Puzzling indeed.

In terms of Carolina bay formation mechanisms, it would seem one is forced to choose between the idea of a wind/water/erosive formation process that respects one of the long-held tenets of the geological community, which is that the morphology of the planet is largely due to slow processes that evolve over enormous periods of time, or the more high-flying notion that the violent impact of a meteor or its secondary consequences caused these bays. Both choices leave much to be desired in terms of explaining what exists on the ground. The inability of wind-based theories to predict what is observed, followed by the rise and fall of the Firestone theory, leaves us with little to hold onto. Surveying the literature backward from 2016 leaves one with a feeling of dissatisfaction, a sense that nothing had been accomplished and that the scientific community knew no more in 2016 than it did a hundred years earlier.

From my own scientific perspective, I have always had trouble accepting a wind-based explanation for Carolina bay formation. The basis for my objection has to do with one word: symmetry. The boundaries of Carolina bays are symmetrical. If you were to print out a picture of a Carolina bay and then fold the resulting cutout down the long axis, the left and right sides would be mirror images of each other. Similarly, if you were to fold the cutout down the short axis of the bay, again, the two sides would be mirror images of each other. This twofold symmetry is simply not observed in objects eroded by wind or water. This is because fluid flow, be it water or air, is an inherently asymmetric thing (except at extremely small flow velocities such as those in microscopic situations), making it unlikely that the type of results shown in Figure 2 would occur.

No matter how many times I look at satellite views of Carolina bays, it is their twofold symmetry that leaps out at me. These lakes (or their outlines once they have been drained) often have boundaries that are beautiful ellipses, quite unlike lake boundaries anywhere else. Indeed, the oriented lakes found in the Alaskan tundra, which we will discuss in the next chapter and which are indeed the result of wind forces, serves as a strong counterexample. Yet many researchers consider these lakes and the Carolina bays as the same thing. Images of the Alaskan oriented lakes are often described as oval. And when you look at them, you might agree. Kind of. Sort of. Maybe. Certainly they all look elongated, and occasionally you might find one that looks oval. But the overwhelming majority have a portion of their boundary that spoils the ellipse—it is flat or rough, yielding a lake that looks more like a bullet or a teardrop than an ellipse. Indeed, one of the classic papers on oriented lakes, by Robert F. Black and William L. Barksdale of the U.S. Geological Survey, describes them variously, as, "...elliptical, cigar-shaped, rectangular, ovoid, triangular, irregular, or compound."

Carolina bays on the other hand are close to perfect Euclidean shapes. The physical world rarely provides perfect Euclidean shapes. Search the rocks found on any piece of ground near you and you will likely find rough, irregular shapes. If you do happen to find some beautiful river stones, the shapes will be smooth, but they won't be perfectly spherical or elliptical. (Indeed, so rare are round stones that the Pueblo Indians of the Southwest collected them and consider them sacred.)

We know that erosive processes smooth shapes. An exposed rock becomes softer in shape as time passes. Sand grains in the ocean become rounded. But rounded and round are two very different things, and even the most casual view of a handful of sand reveals a distinct lack of spherical objects. An earthquake creates irregular cracks in the earth. Coastlines look like dendritic scribbles, mountain ranges are jagged, islands are never round, rivers are never perfectly s-shaped, and lakes are never perfect ellipses...except for the Carolina bays.

Living things produce Euclidean shapes all the time. The pupils of our eyes are circles and, when unstressed, our red blood cells are discs. Microscopic images of pollen reveal myriad regular shapes both complex and beautiful. The components of compound eyes in insects can be hexagonal or square. But living things are, well, living. The Euclidean shapes we see in the world of biology exist because such geometries provided some kind of evolutionary advantage. But the shapes we see in the physical world did not evolve. The shapes we observe in the physical world are determined by the equations that govern physical processes like fluid flow and erosion.

An exception to the above is the physical process of impacts. Impacts *do* result in symmetry, as craters on both the Moon and Earth affirm. We also know that physics permits an ellipsoidal shape from low-angle impacts, though the conditions for this to happen are very rare. If there were only a handful of Carolina bays in the world, an impact theory might be appropriate, but there are hundreds of thousands of them. Up until 2017, there seemed to be no way to reconcile the stringent conditions required for an elliptical impact with the large number of Carolina bays. An impact explanation seemed both possible and impossible at the same time. It was as if some tiny piece of the puzzle was missing.

And, in 2017, a man named Antonio Zamora published a paper that seemed to provide that missing piece.

According to his website, Antonio Zamora does not have the background typical of the university professors and government scientists who publish in the geology journals where work on the Carolina bays typically appears. He is neither a professor nor a geologist. Instead, he has degrees in chemistry and computer science, and has worked for IBM and as an independent consultant. One would have to characterize him as something of an outsider, and his website (www.scientificpsychic.com), with links to online tarot-card readings and chats among other oddities, does little to create an image of a man committed to the scientific method. Yet despite his outsider status, or perhaps because of it, his paper, "A model for the geomorphology of the Carolina Bays," published in the peer-reviewed journal *Geomorphology*, contains interesting new facets in support of an impact hypothesis for Carolina bay formation. He offers what he calls a Glacier Ice Impact Hypothesis, suggesting the Carolina bays were formed when a meteor struck the Laurentian Ice Sheet over what is now Saginaw Bay, Michigan.

Zamora's hypothesis is initially attractive. He suggests this impact did not result in the ejection of rock, but rather of ice—an enormous number of large ice boulders. These subsequently impacted the Atlantic Coastal Plain, thereby forming the Carolina bays. Zamora makes deft use of existing results of other researchers and also recognizes or emphasizes points others have largely ignored. Zamora notes there is wide variance in the use of the words *oval* and *ellipse* in the literature on Carolina bays (to be fair, a blurring of meaning that exists in this book as well). While an oval lacks a particular mathematical definition, an ellipse has one. Especially important, Zamora notes that an ellipse is the shape that you get when a cone is intersected by a plane. Or, to be more relevant to the subject at hand, when a conically shaped object intersects the ground. Thus, if a conically shaped ice impactor were to strike and penetrate the ground in the Atlantic Coastal Plain, it would form an ellipse, a process shown in Figure 3.

Zamora goes on to take advantage of existing research on the geometrical characteristics of Carolina bays that shows the orientation of the long axis of the bays does not point in exactly one direction but varies as you move up the Atlantic Coastal Plain in such a way that the long axes of the bays intersect roughly over Saginaw Bay. This enables him to identify that area as the possible meteor impact location. He also references earlier work in the literature showing that the short axis to

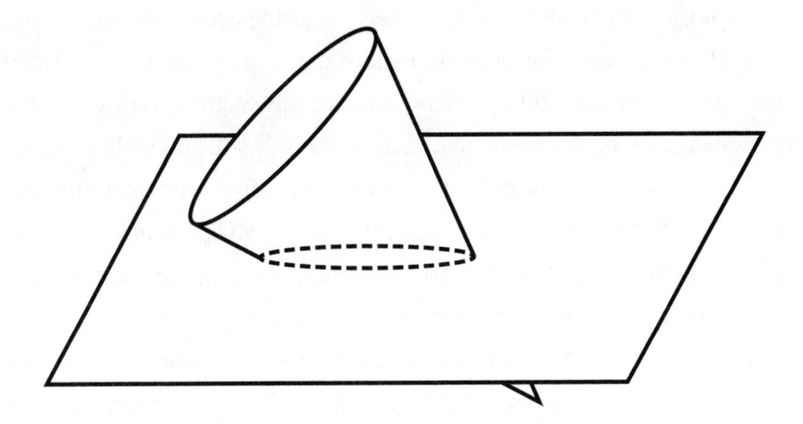

Figure 3. Diagram showing the formation of an ellipse (dashed) when a cone intersects a plane.

long axis ratio of the bays has an average value of 0.58, a value attained when a cone intersects a plane at an angle of 35 degrees, giving the angle of impact for his proposed conical ice impactors. Working backwards from these data, Zamora was able to calculate the numerical characteristics of the impact. The primary impactor would have to have a diameter of 1.9 miles and an impact velocity of almost 11 miles per second. The resulting secondary (ice) impactors would strike the earth in the Atlantic Coastal Plain at a speed of between 1.9 and 2.5 miles per second. They would travel for between six to nine minutes from the impact site before hitting the Atlantic Coastal Plain. For much of that time they would be outside of the Earth's atmosphere, where there would be no melting due to the heating that accompanies such high velocity travel through air. Heating during exit and reentry of the atmosphere would be a factor, of course, but it would melt the smaller boulders. There would be a set of ice boulders large enough to survive.

At this point, a problem with Zamora's theory arises. A velocity of 1.9 to 2.5 miles per second for the conical ice impactors is large enough that an explosion on impact would be likely, resulting in circular impact craters. He anticipates this objection and states that an exception exists if the impactor strikes soil that is undergoing liquefaction. In this context, liquefaction does not mean heating up the sandy soil and melting it, but rather refers to the type of liquefaction that is often seen on mountainside soils during earthquakes. Here, the violent shaking causes the soil particles to move with respect to each other, enabling the soil to flow as if it were a liquid. Zamora claims that liquefaction would occur in the sandy soils of the Atlantic Coastal Plain because the shock waves of the incoming ice impactors and the impacts of early impactors would serve to liquefy the soil in the same way that an earthquake does, enabling the follow-on impactors to penetrate a liquefied soil without an explosion.

The pieces of the theory seem to fit together. But, much like Firestone's work, the theory required the existence of an extraterrestrial impact in the Michigan region. And here is where Zamora's model more or less falls apart: no such evidence seems to exist. There is no evidence of a crater in that region. In 2019, a group of authors from Czechoslovakia led by Jaroslav Klokočník published an article in the *Journal of Great Lakes Research* describing their use of gravitational-field data to suggest possible evidence of an impact in the Saginaw Bay area. However, a commentary article by a five-member team led by Randall J. Schaetzl of Michigan State University challenged the findings, noting the low spatial resolution of the data used and other flaws in the work, ultimately countering that if an impact occurred in Saginaw Bay, no concrete evidence exists. Without evidence of an impact, it is difficult to accept a theory for Carolina bay formation that relies so heavily on such an impact.

As of this writing, there seems to be little else to report from the scientific journal literature on the formation mechanism of Carolina bays. The same cannot be said for the internet, where online forums, blogs, and websites on the Carolina bays can be found in relative abundance. The opinions and ideas evinced on many of these venues seem to be those of a mildly aggrieved community protesting a scientific establishment hostile to any idea that involves an extraterrestrial explanation for Carolina bay formation. The arguments presented demonstrate time and again the "possibility" of a catastrophic mechanism for bay formation (as well as other unexplained aspects of geology). But much like Zamora's theory, geological evidence of some kind of impactor is needed—and, none is found. Perusing these forums, one gets the sense that the participants are committed to the belief that Carolina bays were formed by some extraterrestrial object and are doggedly looking for supportive evidence. In other words, the approaches taken by these researchers suffer from no small amount of confirmation bias.

None of this is to say that evidence won't one day be found showing the Carolina bays were indeed formed by an extraterrestrial impact. The history of science shows numerous examples of accepted wisdom being thrown on its head. But until then, an adequate explanation for how the Carolina bays were formed does not exist. And maybe that is not such a bad thing. While we all love learning of fascinating new discoveries and explanations for how and why the Earth is the way it is, it is also heartening to still have a little bit of mystery in the world.

5. ORIENTED LAKES

In northern Alaska a line of mountains called the Brooks Range runs roughly east–west and defines the southern boundary of what is called the Arctic Coastal Plain. Here lies an immense swath of tundra, an enormous region some 25,000 square miles in extent, more than four times the size of Connecticut. This is the North Slope of Alaska, an area containing the National Petroleum Reserve and the Arctic National Wildlife Refuge (ANWR), places that, if you have heard of them at all, it is from news stories on the battle between environmentalists and the petroleum industry over the region. It is a region with little topography. Once you move north from the foothills of the Brooks Range, the land is quite flat sloping only gradually as you travel toward the Arctic Ocean, to the Beaufort and Chukchi Seas. It is a desolate place devoid of trees other than small dwarf willows. It is also a cold place. This is where you will find Prudhoe Bay as well as Point Barrow, the northernmost point in the United States, where the annual average temperature is around 10 degrees F. Even during the peak of summer, the temperature rarely rises above the 40s. But if you look at this area using the satellite view of a mapping program what stands out has nothing to do with oil, or tundra, or cold. The thing that you focus on is the lakes, for this entire area is simply covered with them. It is a lake district like few others in the world.

On a map where land is color-coded white and water is color-coded blue, this part of Alaska looks like a sheet of paper perforated by a shotgun blast and then held up to the sky: you see blue everywhere. Lakes are the dominant feature of this landscape, and they exist in roughly oval shapes, covering what seems to be more of the region than the land itself. These lakes, referred to as "the oriented lakes of Alaska," are elongated, sometimes two to three times longer than they are wide, sometimes much more than this. All are pointed in a roughly north-south fashion, the major axis of the lakes oriented on average about 10 degrees west of true north.

Studying the map brings forth a curiosity. The prevailing winds in this area are from the east, just a few degrees north of true east, essentially perpendicular to the orientation of the lakes. It is this fact, this 90-degree difference between the wind direction and the lake orientation, that has occupied the minds of limnologists for some time. How can this be, why is it so consistent, what is going on?

It's easy to see why these lakes would attract the attention of a scientist. The satellite views are stunning, and so are the aerial photographs (which one can find in many of the earlier papers published on these lakes). Some of these papers were published in the late 1940s and early 1950s. That, in itself, is one of the striking things about the oriented lakes of Alaska—how scientific attention was drawn to these difficult-to-access lakes, well before satellite imaging and other advanced remote-imaging methods were available.

Northern Alaska is very remote. Prudhoe Bay is almost 500 miles from Fairbanks and over 800 miles from Anchorage. Even today, with the feverish thirst for oil and natural gas, the largest habitations in this petroleum-rich region are small. According to the 2010 United States census, the population of Prudhoe Bay is 2174 and Point Barrow (recently renamed Utqiagvik) is 4212. These are not easy places to get to. The Dalton Highway, which runs from Anchorage to Deadhorse (just south of Prudhoe Bay), was built in 1974 and is still an almost all-gravel road, used primarily by heavy trucks serving the oil fields. Other than that, access to this area is primarily by air. When research on these oriented lakes began in the late 1940s and early 1950s, the challenges associated with getting to and around in this area were quite significant. Yet studies involving both field campaigns on the ground and aerial photography were conducted then—clearly a response to the intriguing nature of these unique lakes.

As noted in the previous chapter, the Carolina bays are sometimes lumped into the same category as these oriented lakes. From a cursory glance, one might suspect this is indeed the case. But closer inspection of oriented lakes shows significant differences. While the Carolina bays possess a strikingly elliptical outline, examination of any map of northern Alaska shows that the lakes found there exhibit a much wider range of shapes. While virtually all are oblong and many are roughly oval, some verge on the rectangular, while many others have various random asperities or irregularities in their outline. The two long lake shores along the major axis direction are rarely similar in shape, one often a bit more curved than the other. A number of them have a kind of teardrop shape, with either the north or south shore being more pointed than its opposite; in some cases this tendency verges on the extreme, making the lake almost triangular. Moreover, a significant difference between these lakes and the Carolina bays is that the oriented lakes of northern Alaska are still being formed. The land here is permafrost and when these lakes melt during the brief summer, their shapes continue to evolve. (This is not the case for the Carolina bays, which are no longer changing in shape, certainly not at the rate of the oriented lakes of north Alaska.) But, like the Carolina bays, the main question that strikes you when you look at these oriented lakes concerns how they formed. How did they get their consistently oblong shapes? Why are they all pointed in the same direction? Why is that direction perpendicular to the wind direction?

So, let's explore what is known about how these lakes formed and are being formed. Let's find out how these oriented lakes were born. Along the way, let's also examine the progression of ideas regarding their formation. In other words, instead of just looking at the most recent theory for how oriented lakes form, let's follow how we got to our current understanding and, in so doing, learn something about how scientific ideas evolve.

Science is always in a state of flux. The ideas surrounding any interesting question experience continual challenges from new thinkers who postulate alternative theories for why things are the way they are. This process is often masked from public view. In a world increasingly affected by science and technology, the news often informs us of what "science says," or what "scientists say." This view implies a degree of agreement in the scientific community that is often at odds with reality, where disagreement and outright conflict of views is not uncommon. We are all familiar with such disagreement in a historical sense; for instance, we've been taught how our understanding of astronomy evolved from an earth-centered solar system to a sun-centered one. But such events always seem to be communicated as stories of a different era—a time when we didn't understand how things really worked—a time before we uncovered the truth.

But in virtually any field, we really don't know how things work and are in a continuous search of truth. That continuous process was readily visible in studies of the Carolina bays described in the last chapter. But those studies were exposed to the glaring hot lights of the popular press amidst something that approached a scandal, a rare situation. The investigations of the oriented lakes of Alaska never excited such attention, and as such enable us to view the growth of scientific understanding under conditions more typical of scientific research.

• • •

Given that the oriented lakes of Alaska are in the process of forming, one might expect the story of how these lakes formed would be a much more mature one—a story with widespread agreement. These are, after all, lakes whose changes can be documented today without resort to the fossil record or indirect markers of a putative formation mechanism. One can go to an oriented lake and measure its size, the velocity and direction of the wind blowing over its surface, and the currents in its basin. One can measure how its shoreline is changing. Surely, therefore, the question of the mechanism of oriented lake formation should have been put to bed. But, though there is reasonable degree of agreement on the formation mechanism of oriented lakes, there are still aspects that are incompletely understood. Let's start at the beginning.

One of the very first explorations of oriented-lake formation mechanisms was largely descriptive, published by U.S. Geological Survey employees Black and Barksdale in *The Journal of Geology* back in 1949. Their paper catalogued the topography, drainage patterns, climate, and other facets of the northern Alaska lake district. Included in their paper were numerous aerial photographs and a gorgeously detailed black and white map of the area. It is clear that they had not intended their paper to focus in any way on formation mechanisms. Nevertheless, they included, in what seems almost like an afterthought, a statement that the orientation of the lakes had to coincide with the prevailing winds. The thinking was simply that shores were likely to erode along the direction that the wind blows, causing the lakes to get longer in that same direction. Noting that the current prevailing winds were in fact perpendicular to the long axis of the lakes, they concluded that these lakes were formed during a geologically earlier period of time when the winds were very different from today, specifically during the Pleistocene epoch.

Black and Barksdale were wrong both in terms of when and how these lakes were formed, errors that were corrected by several authors in subsequent papers. This is a shame given that the majority of Black and Barksdale's work was done very well and had little to do with formation mechanisms in the first place. Nevertheless, in subsequent studies their work is discussed in the context of formation theories, focusing on how and why theirs was wrong. For example, Daniel A. Livingstone published an article in the *American Journal of Science* in 1954, based on his Yale University doctoral work, where he notes that Black and Barksdale's suggestion that the lakes were formed during the Pleistocene was wrong. Livingstone found evidence of significant change in the very recent past. Specifically, he presented data related to the age of shrub willows along the banks of elongated lakes that reveal that their shores were moving at a rate of approximately three feet per year. It stood to reason, he argued, that whatever shape these lakes have is the result of ongoing processes—ones that continue to the present day.

To account for the somewhat counterintuitive notion that lakes should erode and elongate in a direction perpendicular to the wind direction, Livingstone then offered a simple but alluring theory. When wind blows over the surface of a lake, it drags water along with it, causing the water to pile up on the downwind (leeward) side of the lake, a process sometimes referred to as *setup* by limnologists and oceanographers. Referring to Figure 4, this means the wind coming from the east will create a slightly higher surface elevation at location A. With water higher at location A and prevented from traveling back across the lake by the wind, it will naturally tend to flow away from point A and toward point B, which has zero setup, since there is no distance (or *fetch*) over which the wind has traveled.

Livingstone developed a mathematical model to describe this process. He included terms to account for the diminishing fetch as one moves laterally away from the centerline of the lake, where the wind opposes the flow as it curves back into the wind. The net result of these two counteracting forces is a longshore current that increases from a velocity of zero at A to a maximum speed at B. At this location of maximum longshore current, there should also exist a maximum in erosive forces that would serve to remove material from the lake edge by both direct erosion and enhanced melting of the permafrost. Thus the lake would grow into an elongated shape with the large axis oriented perpendicular to the wind speed.

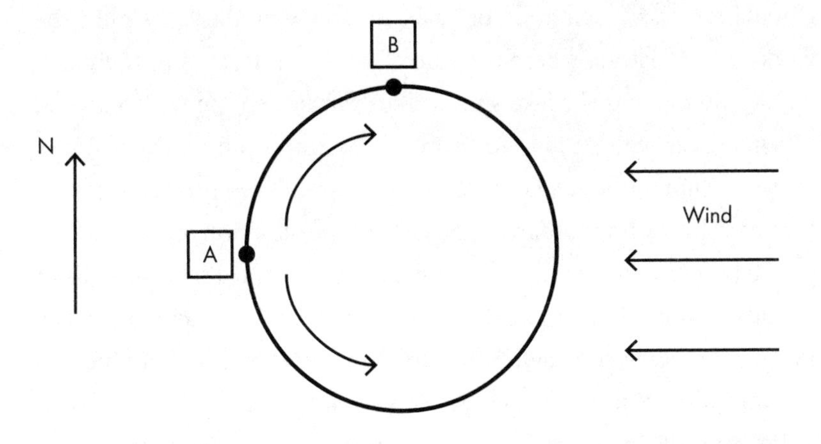

Figure 4. Water current formation mechanism according to Livingstone (1954).

In addition to explaining the shape and orientation of lakes seen in the Alaskan Coastal Plain, this theory of oriented lakes also suggests the existence of two vortices in the lake, a clockwise one at the northern (top) region of the figure, and a counterclockwise one in the southern (bottom) portion of the lake. Thus the wind and surface-water current should travel in the same direction along the short axis of the lake, something that agrees with common sense, but is contradicted by future investigations, as we shall see. Though the existence of vortices was not explicitly noted by Livingstone, in 1957, Hutchinson writes in reference to Livingstone's work that the "...expected circulation consists, in fact, of a pair of antimeric gyrals," gyrals being a less commonly used term for gyres or vortices; antimeric means these two vortices are mirror images of each other.

Though Livingstone's work explained what was observed, it did not remain unchallenged for long. In 1961, a paper appeared in the published proceedings of the First International Symposium on Arctic Geology authored by R. W. Rex and titled, "Hydrodynamic Analysis of

Circulation and Orientation of Lakes in Northern Alaska." Unlike the work of Livingstone where the role of wind is relegated to the pushing of water across the lake surface, Rex focuses on the fact that wind blowing over water tends to form waves. Importantly, when waves impact a shore at an angle, a longshore current (or "littoral drift") is formed. This drift is zero in regions where the direction of wave propagation is normal to the shoreline, for example on the downwind side of the lake, Point A in Figure 4. But, this drift attains its maximum value, Rex shows, when the angle between the normal to the shoreline and the wave direction is 50 degrees. This would be somewhere between Point A and B in Figure 4, but much closer to B. Hence, erosion of sediment is maximized in this region and would cause the lake to elongate in the cross-wind direction, a conclusion also in agreement with observations. This is the same mechanism Kaczorowski refers to; it is described in Figures 1 and 2 of the previous chapter.* Interestingly, however, Rex ends his paper stating:

> *The complexities of single cell versus double cell circulation out in the main body of the lake are not discussed and not necessarily pertinent to the question of lake orientation. The main lake currents are usually much slower than those in the surf zone and are very poorly understood.*

In other words, Rex focuses on waves and specifically on the creation of currents due to waves *at the shore*. Exactly what sort of circulation pattern exists in the lakes as a whole is something he considers unrelated to the evolution of the lake outline. Indeed in the work of Rex, there is no diagram showing how the lake currents are supposed

---

* Kaczorowski considers the angle where maximum erosion takes place to be 45 degrees and Rex states it is 50 degrees. The reason for this goes beyond the scope of this book and is not critical to the ideas we consider here.

to move. Although Rex presents a great many plots and charts and calculations to show the *magnitude* of these longshore currents, nowhere does he indicate the *direction*. This is noteworthy because that direction would be from point B to A in Figure 4, which is to say, *opposite* the direction predicted by Livingstone. This should, in turn, result in a counterclockwise circulation in the north end, though Rex does not say this explicitly. While someone with an understanding of the hydrodynamics of waves and longshore currents would know this, one still wonders why Rex wouldn't have addressed this in a paper 23 single-spaced pages long and where the minutest of details related to the wave propagation and erosion potential were included. Rex's reference to a single-cell circulation is also unexplained, since there does not seem to be a theory for oriented-lake formation that involves a single vortex or gyre.

Setting aside the unresolved issue of Rex's reference to single-cell circulation, the problem seems to boil down to a setup theory resulting in a pair of vortices where the northern vortex is in the clockwise direction and a wave-based theory resulting in a counterclockwise vortex in the north. It would seem that this is precisely the point in the development of an idea where some actual measurements on the ground would come in handy. A simple set of current measurements in these lakes (if traveling by float plane to northern Alaska to conduct fieldwork can be called "simple") would seem to be necessary. And, indeed, in 1959, Charles E. Carson and Keith M. Hussey of the Iowa State University Department of Geology presented such measurements in the *Proceedings of the Iowa Academy of Science*. Their paper contained a detailed description of possible hypotheses for the oriented lakes of northern Alaska, as well as preliminary measurements on surface currents in these lakes. Importantly, the authors claimed a wind-generated current like that proposed by Livingstone and Rex cannot account for the elongation of lakes. Specifically, they claim that the majority of the water motion in the lake was just the oscillatory motion due to wave action.

One gets the impression that in their 1959 paper, Carson and Hussey got it wrong. First, by lumping Livingstone and Rex in the same category, they seem to have not really understood the differences in the proposed mechanisms of these two earlier authors. Also, in their follow-on paper published in 1960 where they present the measurements of their field campaign in greater detail, they change their story quite a bit. In their 1960 paper, they claim Livingstone's theory doesn't hold because the current measurements they obtained during their field campaign do not support it. However, they now agree with Rex and claim the end currents on these oriented lakes are due to a longshore component caused by waves approaching the shore at an angle and that when that angle is near 50 degrees, this longshore current is maximized. Their current measurements agree with this, revealing a counterclockwise gyre at the northern end of the lake (for a wind coming from the east). They show these circulation patterns in diagrams that illustrate the lake outline, with arrows denoting wind and current directions and showing a counterclockwise vortex in the north. Hence, they disagree with Livingstone and the clockwise gyre in the north.

After their initial publication, Carson and Hussey pressed on. Two years later, they refined their theory and focused on specific lakes in detail. They explored the role of drifted peat on the upwind and downwind shores (the eastern and western shores, respectively). In particular, they looked at how peat protects permafrost from melting, thereby further enhancing the north-south growth of the lake, particularly for smaller lakes, where currents may not be strong enough to cause erosion. For the largest of lakes they again claim elongation is due to erosion from longshore currents caused by waves striking the shore at angles of around 50 degrees—a stance in line with the ideas of Rex. They also seem to have thrown Livingstone a bone by acknowledging that for very small lakes (lakes having a width less than 200 feet), his theory of wind-based setup would dominate, since waves would have insufficient fetch to form on such a small lake.

Close to 40 years later, Carson published a summary. "The oriented thaw lakes: A retrospective," recapped the work of his earlier years (Carson was a student of Hussey during these seminal studies), with no substantial changes or significantly new insights. Namely, for small lakes, yes, wind blows water to the downwind side of the lake, elevating the water level there slightly and causing a flow out toward the north and south ends of the lake. For bigger lakes, the wind creates waves, and these create longshore currents downwind in the downwind corners of the lake and thus create circulation patterns that are counterclockwise in the north.

The fact Carson's abovementioned paper used the word "retrospective" in its title is telling. It implies that his published ideas were now part of the canon, and that only a bit of tidying up of small details was needed. One might think this is the end of the story, and for many years it was. More articles followed on oriented lakes, but none opposed the general point of view that these lakes were formed by waves causing longshore currents. This is a common pattern in scientific endeavor—there is a period of focused research after which a consensus is attained.

But it is often the case that after a pause new investigators revisit the problem with new points of view. This was the case for investigation of oriented lakes when in 2014 (13 years after Carson's retrospective article), Shengan Zhan and co-authors at the University of Cincinnati and the U.S.G.S. published an article in the journal *Remote Sensing*. This was an odd paper because the authors nominally seemed to agree with Carson and Hussey, but really, they did not agree. Carson and Hussey explain the vortices in their lakes using the longshore currents generated by the interaction of waves with a shore. The new authors, on the other hand, come up with a completely new explanation. First, they presume that, generally speaking, in these lakes, the water near the shore is shallower than the water near the center, a reasonable presumption. Then they note that if you were to travel along the long axis of an oriented lake, from the lake center to the northern tip, and you were to identify the point where an equal mass of water exists north of that point and south of that point to the center, such a point would be much closer to the lake center than the northern shore. This means there is a larger lake surface area to the north of this special point than to the south. Thus, if a wind is coming from the east, it would exert a shear over both regions, but the region to the north of this special point would have a larger area for the wind to act on than the region to the south. Since both areas have the same mass, the net result would mean the force exerted on the water north of the special point would exceed the force on the water located south of that special point. The end result of this would be a counterclockwise gyre in the northern end of the lake. This is a *result* that agrees with that of Carson and Hussey, but for profoundly different reasons, a fact Zhan and colleagues do not point out. It is almost as if they were trying to slip their idea into the journal literature—seemingly agreeing with the older more established authors, while actually introducing a profoundly new idea.

After this work by Zhan's group, the journal literature halts. This is odd since Zhan's work seems to be a challenge, albeit camouflaged, to the generally held understanding of how oriented lakes form. Of course these authors do not argue against the idea that waves do the *work* of erosion in these oriented lakes, and so it is far from a complete rejection of earlier ideas. For example, it may be that the wave-induced currents proposed by Carson and Hussey do most of the erosion work, while the currents induced by the mechanism of Zhan's group simply contribute to those same flows. Still, one would expect some counterarguments to arise. However, titles that contain "oriented" and "lakes" cease after Zhan's work and more sophisticated searching leads to no further articles.

Summarizing, it seems the mechanism of how oriented lakes form is reasonably well understood. Longshore currents formed by waves cause more erosion along points near the northern and southern shores of the lake; they also result in a counterclockwise vortex and a clockwise vortex in the north and south of the lake, respectively. This causes a growth of these lakes in the crosswind direction. The work of Zhan's group suggest that the wave-induced currents may be enhanced by a different mechanism completely unrelated to waves. Still unresolved is whether the method of current formation proposed may be a significant player in lake erosion or not. Further studies may arise in the future. But, even if the mechanism of Carson and Hussey is the complete story of oriented lake formation, it is still a pretty interesting story. We are used to thinking of wind-induced erosion as something that acts in the direction of the wind. We have all seen images of rocks and trees where their upwind sides have been eroded away, leaving, for example, a tree whose limbs point only in the downwind direction. So, it is odd, almost unnatural, to think of wind causing a lake shore to erode, but in such a way that material is removed primarily in the *crosswind* direction.

Hopefully this chapter has adequately explained how the oriented lakes of northern Alaska are born and grow, revealing how the complex interaction between wind and water flow causes a counterintuitive mechanism for lake formation. This intriguing topic is also an illuminating example of how science proceeds—how the fits and starts of research result in an evolving story of how things work—how the story seems to be complete and then is modified, sometimes turned on its head. In the case of oriented lakes, it seems research is presently at a pause. However, we should not be surprised if tomorrow an alternative explanation is published in the journal literature, one modifying or even completely changing what we once thought was true. This has happened in science many times before and surely will happen many times in the future.

6. SUBGLACIAL LAKES

In 1960, the Soviet pilot R. V. Robinson noticed extensive flat areas in the otherwise wrinkled surface of the Antarctic ice sheet. He called these areas *lakes*, probably never knowing how right he was. Robinson surely knew that these smooth areas were ice, frozen solid by the frigid temperatures. He would have known that these flat surfaces, though they looked similar to the frozen surface of lakes in more temperate climates, did not have liquid water a few feet below. But in a way Robinson never could have known, these regions of flat ice were indeed the covering of a lake, albeit a lake located miles beneath the surface.

Underneath Russian research station Vostok on the East Antarctic Ice Sheet, lays Lake Vostok, a lake unlike any other. No human has yet seen Lake Vostok, which is a pocket of water trapped between the Antarctic bedrock below and the massive Antarctic ice sheet above. It is a lake without an air/water interface. It has a rock/water interface at its bottom and an ice/water interface at its roof, but that roof is over two miles distant from the atmosphere, far from air and from light. Lake Vostok is enormous, its estimated volume close to that of Lake Ontario. And, what is especially interesting from a biological point of view, there is no direct interaction between the water in Lake Vostok and Earth's atmosphere. The water in this lake has not been exposed to air for at least 100,000 years, and by some estimates not for a million or perhaps even ten million years. This means whatever life resided in that pocket of water as the Antarctic ice sheet grew is still present, having evolved independently of the bacteria and fish and mammals located far above. Although researchers speak only of comparing the bacteria from Lake Vostok with those found elsewhere on Earth, there is always the unspoken possibility: What else might live there? Could something larger than bacteria have learned to survive there over the course of a million years? Could evolution have proceeded in these subglacial lakes along a path different from that found in lakes on the surface far above?

Lake Vostok is by far the largest of the known subglacial lakes. It is 155 miles long and about 50 miles wide with a total surface area of 5400 square miles and a volume of 1200 cubic miles. The thick sheet of ice above Vostok is impressive, but Vostok itself is also impressive, having a depth of more than 3280 feet, deeper than several of the very deep lakes we've already discussed such as Crater Lake and Lake Nyos; it is deeper than all of the Great Lakes. Lake Vostok is so big that it ranks as the seventh largest lake in the world.

But while Lake Vostok is the largest subglacial lake on Earth, and the first discovered, it is by no means the only one. Excitement over its discovery spurred studies of subglacial lakes in general. Aided initially by seismic soundings, and then radio soundings, and finally by satellite scans, glaciologists have found many more subglacial lakes, estimates rising steadily from little more than a dozen in the 1970s to about 400 today. Much like their terrestrial counterparts, subglacial lakes are often connected by subglacial river systems, some lakes rising at the expense of others. These transfers of water from one lake to another can be significant enough to change the height of the ice sheet located miles above the water surface. We now know the base of the Antarctic ice sheet has something of a secret hydrological life, where meltwater is channeled along riverine systems to and from various lakes. Indeed, it is estimated that as much as half of the glacial base of Antarctica is wet, not frozen, covered either in lakes or in flows of meltwater traveling from one place to another.

Antarctica is a very cold place, and exactly how any liquid water can exist at all, let alone an enormous body of liquid water like a subglacial lake, is an interesting question in and of itself. The temperature above Vostok Station, the Russian outpost where much of the research on Lake Vostok takes place, is extremely cold with temperatures as low as –75 degrees F. Indeed, the lowest temperature ever measured on Earth was recorded at Vostok Station, on July of 1983: –128.6 degrees F. A temperature this low is something to ponder. If you were in a place where the temperature was that low, and you were somehow able to increase that temperature by a whopping 100 degrees F., you would still be just shy of 30 degrees below zero, requiring yet a second 100-degree boost in temperature just to get to room temperature. If you wanted to melt and then boil a piece of ice from the ice sheet at Vostok on that very cold day in 1983, you would have to increase its temperature by 341 degrees F. Vostok is clearly a place that does not favor the liquid phase of water.

It is disconcerting to realize such a harsh environment exists on our planet, a place so cold that the liquid in our own bodies would always be on the verge of solidification. Yet, in spite of this incredible cold, liquid water does exist in Antarctica, albeit far from the surface.

The reason this is possible involves several factors. First, when we speak of the incredible cold in Antarctica, we are referring to the air temperature. The air temperatures at the poles are cold due to the reduced energy received from the sun. Indeed the South Pole is dark 24 hours a day during its winter, receiving virtually no energy from the sun during this time. In the summer it is light all day long, although the altitude of the Sun barely skirts above the horizon, reaching its peak zenith at a measly 23.5 degrees above the horizon at the summer solstice, providing very little energy to the surface. In contrast, regions closer to the equator receive significant energy from the sun. In temperate regions, digging a hole in the ground generally exposes soil that is cooler than the air above it, certainly in the summer, something your dog may have taken advantage of to cool her belly on a hot summer day. But at the poles, drilling downward into the ice sheet actually results in an increase in temperature.

To better understand this, it bears remembering that the center of our planet is quite hot, and the heat generated in the core conducts outward to the surface. The amount of heat emanating from the Earth's surface due to the hot core of the planet is called the "geothermal heat flux," and its value varies with location but is typically around 50 milliwatts per square meter. This is not a very large flux of heat, and it has little effect on the temperature near the equator or temperate regions of the planet where the flux of heat from the Sun exceeds 1000 watts per square meter (one million milliwatts per square meter), over 20,000 times the geothermal heat flux. Indeed, if one sought to obtain the amount of energy needed to power an old-fashioned 100-watt lightbulb from this geothermal heat flux, a surface area of over 21,500 square feet would be needed, a patch of land 147 feet on a side. But, though small, this quantity of heat can melt ice at the bottom of the Antarctic ice sheet because of several factors. First, at Vostok, the ice sheet is approximately 2.5 miles thick. Much like the much thinner walls of an igloo, this blanket of ice serves effectively as an insulator, protecting the bedrock from the very cold air at the surface. So, although the geothermal heat flux is quite small, the energy it provides from the Earth's inner core is not snatched away by the howling winds of an Antarctic winter but is held in place, much like the heat emanating from your body is kept in place by a good coat.

Secondly, as we learned in an earlier chapter, glacial ice sheets are virtually always in motion, continuously sliding downhill from higher regions of ice accumulation. This flow results in frictional heating as the ice slides over the bedrock or flows by deformation under the enormous pressure of the ice sheet. Such heating adds to the geothermal heat flux, providing another source of energy to melt ice. Finally, the immense pressure at the bottom of the ice sheet reduces the melting point of ice, decreasing it by 5.4 degrees F., from 32 degrees F. to 26.6 degrees F., allowing the ice to melt at a lower temperature than would otherwise be the case. Combined, these factors cause ice to melt in the region of

Lake Vostok, as well as in the many other regions beneath the Antarctic ice sheet where subglacial lakes exist.

Given that water can melt in the region of Lake Vostok, one might wonder what keeps the process in check? Why isn't Vostok even bigger, with more of the ice sheet melting to create more lake water? Or, what keeps it from shrinking, from cooling just enough to reverse the melting and cause the lake to solidify? The answer comes from an energy budget for the lake, a comparison of the amount of heat coming in compared to the amount going out. Heat comes in from geothermal heating and from frictional heating and perhaps from relatively warmer meltwater flowing in. Heat is lost if relatively warm meltwater flows out of Vostok*, and heat is lost to the colder ice that slowly flows across the lake surface, sucking heat away from the relatively warmer lake water. If the amount of heat entering the lake exceeds that lost, melting will occur. If the amount of heat leaving exceeds that going in, it will cool and perhaps freeze. The lake size will grow or shrink until equilibrium is attained, a point where the inflows and outflows of energy equal each other, at which point the lake size will be stable.

But the above treats all of Lake Vostok and the other subglacial lakes as if they are homogenous, each part of the entire lake under more or less the same conditions. But the situation is far more complex and interesting. Indeed, along the roof of Lake Vostok, freezing is actually occurring in some places, ice accreting in that locale, while melting occurs in other locations where ice is lost. This process leads to a circulation of water within the lake, hardly a torrential flow, but vigorous enough to suspend sediment from the lakebed into the water. This is interesting because it means that the water samples that have been obtained from the lake (more on which below) could reveal its history. The suspended bottom sediment may provide a window into the ancient life that has

---

* It is unclear if there actually is a flow of meltwater in/out of Vostok. But this process does occur for other subglacial lakes.

existed in this lake. The flow is due to a complicated chain of physical events, and it all depends on a small bit of ice/water physics, the simple but odd fact that the roofs of all subglacial lakes are tilted.

At its northern shore, the roof of Lake Vostok is about 1310 feet lower than at its southern shore. It is a strange fact, and one with profound consequences. It is strange because at the surface of the ice sheet above Vostok, there is an opposing and much smaller tilt of just 131 feet. The situation is revealed in Figure 5, which shows first the situation one might expect to exist in a lake covered with ice and, in the lower part of the figure, the situation that actually exists at Vostok as well as at other subglacial lakes (albeit with different thicknesses).

The fact that the surface of the ice is tilted one way and the roof of the lake is tilted the other is counterintuitive and probably best understood by focusing on the floor of the lake. Let us imagine this floor is flat as shown in Figure 5 (it is not, but this does not affect the mechanism we are trying to understand). Now, referring to the upper portion of the figure, let's begin with the simple case where the ice sheet is flat at the surface. The figure shows two locations, A and B, both at the bottom of the lake. In the upper portion of Figure 5, the combined weights of the water and ice are the same at A and B because the thickness of the ice and water layers above each of these points is the same. Let us now imagine the amount of ice on the surface changes so that there is 131 feet more ice above point A than above point B. In other words, imagine the upper ice surface is tilted so that it is 131 feet higher on the left than the right. This would make the weight at A higher than at B, therefore the water pressure at A would be higher than B. Such a situation could not exist in equilibrium, and the system would adjust itself so that equilibrium is reestablished. Here, the ice/water interface tilts until the pressures at A and B equilibrate. Since, in our thought experiment, the weight above A is initially higher than at B, and since ice is less dense than water, the ice/water interface would tilt so there would be more (lighter) ice over A and less water.

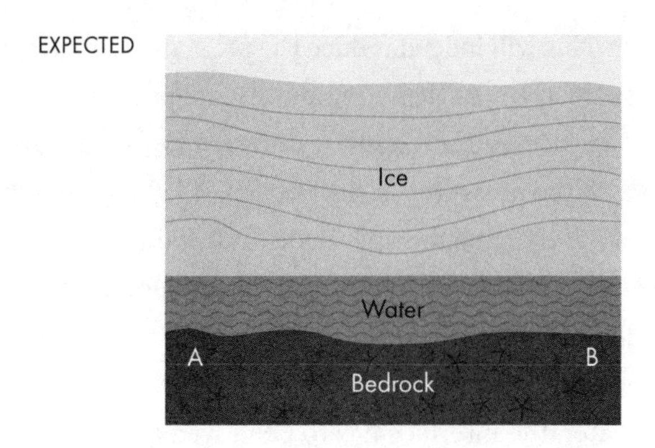

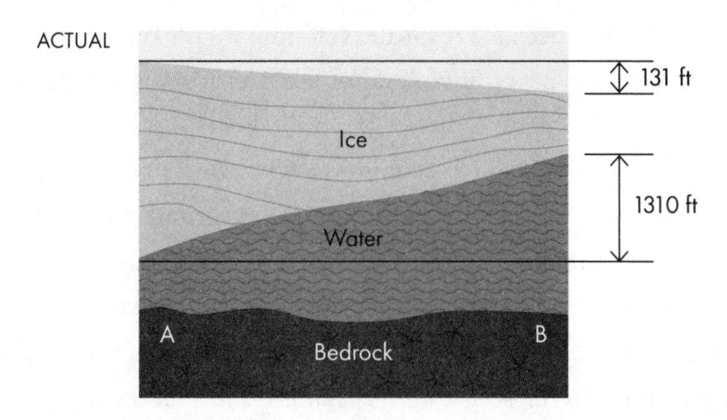

Figure 5. Relative thicknesses of ice and water in a subglacial lake. Above is the situation one might expect, and below is what actually exists in Lake Vostok and likely in all subglacial lakes. A is the northern shore of Vostok and B is its southern shore.

This probably makes a reasonable amount of sense. Where the situation becomes harder to understand is, exactly why the small tilt of the upper ice surface (131 feet) must be compensated by such a large tilt of the ice sheet at the ice/water interface (1310 feet). Why this ten-to-one ratio? The answer takes a bit a mental focus but can be understood if one imagines slowly pulling down the left side of the line defining the ice/water interface. If you pull that line down on the left side of the diagram, you are increasing the thickness of ice and decreasing the thickness of water. This will indeed reduce the pressure at A, bringing it closer to the pressure at B—so that equilibrium is approached.

However, though ice and water have different densities, they are not *that* different. At the pressures of the bottom of Lake Vostok, liquid water has a density of 63.43 pounds per cubic foot, while ice has a density of 57 pounds per cubic foot, a difference of only about 10 percent. Hence, pulling down the left side of the ice/water interface does add lighter ice and reduces heavier water, but since the ice is not *that* much lighter than the water, the interface has to be moved quite a bit before the water pressure at A and B are in equilibrium, as they must be. This is especially true because regardless of how we tilt the ice/water interface, the total height of liquid and/or solid above A is still higher than that at B.

It is a difficult bit of reasoning to wrap one's head around. But, regardless of the reason, the tilt in the ice/water interface is about ten times larger than at the surface of the ice sheet. This is the case for all subglacial lakes. The relatively small slope on the surface of the ice sheet at Vostok causes the roof of the lake to change by about 1310 feet from the north to the south shore. That is quite a bit of change, considering that Vostok is about 3280 feet deep at its deepest location.

The change in height along the roof of a subglacial lake has significant consequences. It is widely believed that this slope is responsible for circulation in these lakes. The temperature of liquid water immediately adjacent to an ice/water interface should be quite close to the freezing point. In Lake Vostok, the northern end of the lake is 1310 feet deeper than the southern end and due to the different pressures at those depths, the freezing point is therefore about 0.5 degrees F. lower at the northern end than the southern end. Hence, the northern water near the roof is colder and denser. This denser water sinks and partially follows the bottom of the lake, traveling to the south. The interesting thing is that this cooler water is being transported to the southern end of the lake where the freezing point is higher, causing that water to freeze. Hence, ice is accreting at that end. To maintain an ice and heat balance, there would have to be melting and the formation of meltwater at the northern end of the lake. Radar soundings show this is precisely the case.

The fascinating thing about all of this is that this circulation mechanism applies not just to Lake Vostok, but to all subglacial lakes. All subglacial lakes are believed to have sloped roofs and to experience this type of circulation driven by a difference in melting points between the thick-ice end of the lake and the thin-ice end.

Now, the time scales of all of the processes described here are slow—glacial. The ice flow over the surface of Vostok is estimated to move at about 10 feet per year, taking tens of thousands of years to flow over the entire lake surface. The rate of ice accretion at the locations where water freezes is about an inch per year. And though all of these rates pale in comparison to water velocities in a temperate lake (where water velocities can be orders of magnitude larger due to wind and inflows and rain), there is still just enough vigor in these flows to enable sediment at the bottom of Vostok to be suspended in the lake water. The lake water velocities are estimated to be just strong enough to keep particles as large as 20 microns in diameter suspended in the water. Indeed, Vostok may be turbid.

Because bacteria are typically much smaller than 20 microns, it is believed that water samples from this mysterious, prehistoric Lake Vostok, should reveal not just what kind of life exists in the lake now, but via sediment particles, should also reveal the history of the life that existed in its prehistoric past. To find out, scientists would need to get a sample, which would require drilling through 2.5 miles of ice, without introducing bacteria from the surface, bacteria from the modern era, into this ancient lake. Accomplishing this was not without its challenges, but it has been done.

To date, drilling into subglacial lakes in Antarctica has been pursued at three locations. Drilling at Lake Vostok, a primarily Russian activity with support from other nations, has been ongoing. A U.K.-led effort exists at Lake Ellsworth, a narrow subglacial lake about 8.7 miles long, with a width mostly less than a mile, a maximum depth of 512 feet and a volume of about 0.3 cubic miles. It is covered by an ice sheet about 2 miles in thickness. A U.S.-led effort exists at Lake Whillans, which has the advantage of having an ice sheet "only" about 2600 feet thick. Whillans is thought to be relatively shallow, perhaps only a few yards deep, and also very hydrologically active; two periods have been recorded when the surface ice above Whillans has risen and fallen. As of 2017, of these three opportunities, two have resulted in the penetration of the ice core into the lake and have enabled retrieval of samples from the lake. These were Lakes Vostok and Whillans.

When the Russian team broke into Lake Vostok, they did so via a borehole they'd been developing for over 15 years. That borehole was initially used to obtain an ice core with an end point that came within about 500 feet of the water surface and yielded useful information in and of itself. But in February 2012, the team broke into Lake Vostok proper. When the drill bit penetrated the surface, lake water rushed upward and promptly froze. The next season, a core was taken of this lake water, referred to as *fresh frozen* lake water, and was made available to the scientific community. The results showed the existence of life, albeit all microbial in nature, and analyses of the bacteria and fungi in this core and in similarly obtained cores has been ongoing.

But the Russian program had always been somewhat controversial.

Drilling in Antarctica, particularly eastern Antarctica, which is colder than its western counterpart, is challenging to put it mildly. One of the difficulties with such an endeavor is simply ensuring your drill bit does not freeze to the bore hole, not an easy feat at air temperatures near –100 degrees F. One way to prevent freezing is to utilize a drilling fluid in the borehole that does not freeze. This was the approach that the Russians took, utilizing a mixture of kerosene and freon which stays liquid even at the extreme temperatures of eastern Antarctica.

But kerosene is not as antiseptic as one might think; strains of bacteria call it home. So, by the time the drill bit broke through Vostok the miles-deep hole was filled with an estimated 60 tons of kerosene/freon mixture. A very real concern was that some of this mixture would fall into the lake when the bit penetrated the roof, forever contaminating the prehistoric lake with modern bacterial strains. But, because kerosene/freon is of lower density than water, when the lake surface was penetrated, lake water rushed upward. Thus the drilling fluid did not fall into the lake. From the perspective of preventing contamination of the lake, this was a plus. But at the same time, the ice cores obtained from the Lake Vostok borehole were contaminated, or such bacterial contamination could not be ruled out. Hence, although the core from Lake Vostok revealed the presence of microbial life, the significance and accuracy of these results was questioned.

A way to avoid the contamination of the Vostok approach is to use a hot-water drill. This technique prevents the drill from freezing to the ice, while also ensuring that the borehole is clean by implementing a sterilization procedure. Such an approach would ensure no external bacteria are introduced into any samples obtained from the lake. This approach was taken at Lake Ellsworth but failed due to equipment malfunctions. However, when a similar approach was applied in 2013 at Lake Whillans, it was successful. Lake samples from that work revealed a system of microorganisms that utilize chemosynthesis to extract energy from the minerals present in the lake.

In both of the subglacial lakes penetrated in Antarctica, evidence of anything larger than microorganisms has not been obtained. Moreover, truth be told, none of these drilling activities were pursued with the hope of finding evidence of aquatic dinosaurs swimming about. The expectation was for evidence of only microbial life. Bear in mind that these lakes are all completely dark, precluding the possibility of photosynthetic life. Furthermore, carbon sources in subglacial lakes are limited to organic matter scraped into the lake by the ice sheet flowing over it, or whatever may enter the lake from the ice above, that is, matter deposited onto the ice sheet that slowly made its way downward as the ice sheet formed over the interceding millennia. This is not a very large flux of carbonaceous matter and therefore sources of nutrients to sustain any life that might exist in Vostok are thought to be limited. There are also sediments that may have been left behind since before the Antarctic ice sheet existed, at the very beginning of the current ice age millions of years ago. But, by and large, whatever life exists in Lake Vostok would have to make do with minimal sources of energy and nutrients.

Nevertheless, a full understanding of the ecology of Lakes Vostok and Whillans will likely require further drilling, particularly into the floor of the lakes to further understand the sediments at the bottom. How this may be accomplished and what it might reveal are still to be determined. Even if these two lakes are found to contain nothing of especial interest, the story is unlikely to end there. It is estimated that there are more than 400 subglacial lakes, and there is nothing to say the ecosystems of these lakes have to be the same. Meanwhile, on the other side of the planet, a subglacial lake was identified in 2018 in the Canadian Arctic. This time the lake is a salt lake, suggesting that the story of subglacial lakes on Earth is far from over. Moreover, the discovery of subglacial lakes on other planets seems to be just beginning. Liquid water was detected beneath the ice near Mars' South Pole in 2018. This is thought to be a salt lake too, though likely its salts are perchlorates, different from typical salt lakes on Earth.

Clearly, we are living in the salad days of subglacial lake research, likely with many revelations in the future. There has been absolutely no evidence to suggest the existence of dinosaurs in the subglacial lakes of our planet. But still, I keep my fingers crossed.

7. SALT LAKES

Compared to their freshwater counterparts, salt lakes get little attention. Perhaps this is because they contain water that cannot be used for irrigation or drinking. Perhaps it is because they tend not to be home to abundant fish populations (although there are exceptions to this). Maybe it is simply because we expect salt water to exist only in oceans and its presence inland seems odd to us. But whatever the reasons salt lakes are ignored, it certainly isn't because of a lack of them.

Salt lakes comprise almost as much of the planet's lake water as do freshwater lakes, a fact that may come as a surprise. Our planet has 24,950 cubic miles of water in salt lakes compared to 30,000 cubic miles for freshwater. Large saline lakes represent 23 percent of the area of all lakes on Earth. Indeed, the largest lake in the world by both area and volume is a salt lake, the Caspian Sea. Clearly, when it comes to lakes, salt lakes seem pretty important. Accordingly, one might expect that, though ignored by the layperson, salt lakes are the object of significant study by limnologists. But this is not the case. Indeed, many classic limnology textbooks pay scant attention to these significant lakes. A classic 1952 textbook on limnology by Paul S. Welch does not have the word *salt* or *saline* in the index at all. An oft-cited limnology textbook from 1975, by Robert G. Wetzel, does include a 24-page chapter on the salinity of inland waters, but this is in a 743-page book, and the chapter only discusses salinity in the sense that all lakes have a finite salinity; it is really more concerned with freshwater lakes than salt lakes per se. Indeed, even the encyclopedic *A Treatise on Limnology, Volume 1*, by G. Evelyn

Hutchinson, a true doorstop of a book weighing in at 1015 pages, lists only two pages in the index that mention salinity. Salt lakes are the poor stepchild of the limnology community and have really only garnered serious attention in recent years.

Salt lakes exist everywhere on our planet, present on every continent, including Antarctica. They can be found at the hottest and the coldest places on Earth, and from the highest to the lowest altitudes. Indeed, the lowest place on Earth is the Dead Sea, a salt lake. Many of the highest lakes on Earth are salt lakes, located in places like Tibet and in the South American Altiplano. Of particular note on the Tibetan Plateau is Namtso, a large lake with an area of over 770 square miles at an altitude of 15,480 feet, the highest known salt lake in the world. Salt lakes exhibit behaviors that are as interesting and, in many cases, far more so than their freshwater counterparts.

Perhaps a good place to start in our investigation of salt lakes is with a definition for salinity, which is typically measured in grams of salt per liter of water. For example, the ocean has an average salinity of 35 grams per liter. Another common measurement is parts per thousand, or ppt, which is equivalent to grams per liter for salt; the ocean also has an average salinity of 35 ppt. Less commonly used is the old-fashioned percent, which is equivalent to parts per hundred; the ocean has a salinity of 3.5 percent, one tenth of its value in ppt. It is also useful to have a definition for the value of salinity that defines the transition from freshwater to saline. In truth, there is a continuum in salinities in the planet's waters, making any boundary between what is salty and what is fresh necessarily subjective. However, a boundary of 3 ppt is widely accepted and has a certain convenience since it turns out to be the salinity at which most humans begin to taste the presence of salt in water.

A good next step in our discussion of salt lakes would be an explanation for why some lakes are salty and others are not. In almost all cases, salt lakes exist because they form in a specific type of basin, termed an endorheic basin. Simply stated, endorheic basins have no outlet. They lack a stream or river to transfer water to a lower lake or to the ocean because the terrain around the lake precludes such a flow or, for the specific case of the Dead Sea, because there simply is nothing lower than that basin. Water comes in via rain and from streams and rivers above the basin. But no water flows out. As far as *liquid* water is concerned, an endorheic basin is the end of the line, a description that justifies the other name used for these lakes: terminal lakes.

The above describes the hydrology of a salt lake but says nothing about the salt's origins. Generally speaking, salt arrives in the water flowing into a lake. At first glance, this is somewhat surprising because the streams and rivers that flow into salt lakes are freshwater, no more or less salty than the water that you and I drink from the tap every day. But, there is salt in the water that flows into the typical salt lake, just as there is salt in the tap water that we drink, albeit small. Water is very good at dissolving things, especially salt, and so there is always some salt in a sample of freshwater.

For a non-endorheic basin, the small amount of salt flowing into the lake largely departs at the lake outlets. But for an endorheic basin, there is no outlet, at least not for liquid water. There is, however, a way for water molecules to leave an endorheic basin, and that is via evaporation. And since most salt lakes are found in arid or semi-arid environments, evaporation rates can be significant. Evaporation removes water molecules, and water molecules only—dissolved materials are left in the liquid. Hence, regardless of what is dissolved in the water of a lake, the water molecules leaving via evaporation are pure, devoid of the salts or minerals they leave behind.

Evaporation's ability to precipitate salt is well-understood. In fact, in laboratories this effect is utilized in distillation units, or stills, where tap water is evaporated in a boiler and then converted back to liquid in a condenser. The resulting distilled water is much purer than the tap water that enters the boiler because only water condenses in the condenser and salts and other dissolved minerals are left behind as scale in the boiler. Similarly, in endorheic basins, water molecules leave via evaporation, leaving any dissolved salt behind. And so, day after day, year after year, liquid water flows into salt lakes, along with a little bit of salt. The water that flows in subsequently evaporates (assuming the lake level holds roughly constant), leaving behind a bit of salt. Over the centuries, the basin's salt level rises and rises, resulting in salinities that can be extremely high. In fact, the salinity can reach its saturated value, the value above which the water can no longer dissolve any more salt. When this happens, further evaporation can cause salt to come out of solution, crystalizing to form strange and beautiful formations.

As noted above, the ocean has a salinity of about 35 ppt, which is not particularly high when compared to salt lakes. In contrast, Mono Lake in California has a salinity of 95 ppt. Of course the salinity of any lake varies with rainfall, inflow rates, and evaporation rate, all of which vary with season and from year to year. Utah's Great Salt Lake has an especially variable salinity, ranging from 150 ppt to 280 ppt.

Salt lakes can even have salinities that vary with location within the lake, a situation that is particularly true when the lake has two or more basins separated by shallow barriers, as was the case for the Dead Sea before its southern portion dried completely. The salinity of a lake can vary to an especially large degree with depth, something that can occur for a range of reasons. Sometimes the salinity will be highest near the surface because this is where evaporation occurs, the lost water leaving a saltier mixture behind. However, it may be that the bottom of a lake is much saltier than anywhere else since this is where crystalized salt will ultimately deposit. All of this can be affected by the degree of mixing in the lake. Lakes that become stratified, and therefore do not mix at all, can exhibit much larger differences in salinity with depth than those that overturn, mix, and thereby homogenize salinity.

As a rule, salt lakes on Earth are getting saltier. The freshwater that flows through the rivers and streams that supply salt lakes around the world can be used for many things: irrigation, human consumption, livestock watering, and other purposes. The temptation to divert these streams away from a salt lake toward what seems like a more useful purpose is difficult to resist, and all over the world the result has been salt lakes whose water levels are falling and whose salinities are rising. This is true of the Dead Sea, Mono Lake, the Aral Sea, and many others.

The consequences of these diversions can be disastrous, as we will see in a later chapter, when we look at the Aral Sea in detail. But, another consequence of this is the formation of hypersaline water, water having a salinity close to or even exceeding its saturation value. Such salinities often result in the precipitation of crystalline salt along the lake shore, or even in the lake itself, forming sometimes thick, extensive salt layers on the lake floor. The Dead Sea is an excellent example of such a lake. Diversions of fresh water by Jordan, Israel, and Syria from the rivers supplying the Dead Sea have been ongoing for some time. This has resulted in a steady decline in its level. Indeed, in the 1920s, the level of the Dead Sea was –1286 feet (the Dead Sea is located below sea level thus its altitudes are always negative), while at the beginning of the 21st century its surface was at an elevation of –1355 feet and as of this writing it is about –1414 feet. The Dead Sea originally consisted of two basins, separated by the Lisan Straits, a narrow, shallow portion of the Sea. But in 1978 the Lisan Straits became dry, creating two separate bodies of water, the Northern and Southern Basins of the Dead Sea. The Southern Basin subsequently dried completely and is now used as a salt works; salt water from the Northern Basin is pumped into the Southern, the water is allowed to completely evaporate, and the remaining crystalized salts and minerals are then harvested.

When the salinity of a salt lake gets sufficiently high, the water becomes saturated, meaning it is unable to dissolve any more salt. In this state, any further loss of water via evaporation can cause salt to precipitate out in the form of crystals. Even when a salt lake is not in a saturated condition, such a process may occur at or near the shores;

stranded pools or even large puddles become isolated when the water level drops and may completely evaporate, leaving salt behind in a variety of crystal shapes. These crystals can be subsequently modified if and when the lake level rises and falls again, causing salts to dissolve, and recrystallize along with the additional salt that was added, forming different crystals each time.

But when an entire body of water such as the Dead Sea becomes saturated, odd things can occur. In the Dead Sea in 1979, small halite crystals (halite is pure sodium chloride) were observed at the water surface. This was the first place saturation occurred since the water surface is where evaporation occurs. These falling halite crystals are sometimes called "salt snow" due to the similarity in appearance of these falling salt crystals to their icy crystalline counterparts falling in air. Then, toward the end of 1982, the entire water column of the Dead Sea saturated and halite began to precipitate out in massive quantities at many depths. Simply putting a rope in the water resulted in the formation of half-inch thick clusters of white halite cubes all along its length. At the sea bottom, thick beds of salt crystals were deposited and are now many inches thick. It is estimated that just between 1976 and 1992, 2.6 billion tons of salt crystalized onto the floor of the Dead Sea. The rate that crystals are forming on the floor is estimated to range between 1 and 4 inches per year. The Israeli artist Sigalit Landau has capitalized on the hypersalinity of the Dead Sea by submerging objects in the water and removing them some time later, thoroughly covered in crystals of white salt. Perhaps her best-known work, "Salt Bride," is a sequence of underwater photographs of a long black dress taken over a period of time as the dress transitioned from black to white due to the progressive crystallization of halite on the fabric.

The crystallization of salts in salt lakes is affected by many parameters, not the least of which is the mixture of different kinds of salts present. For it isn't just pure halite that is present in saline lakes, but a plethora of other salts and minerals such as gypsum, borax, calcite, epsomite, and dozens of others in varying quantities. These can all crystalize together or separately. Moreover, trace elements in an otherwise pure salt like halite can affect the structure of the crystal, as can the rate of crystallization. All of these factors combine to create a seemingly infinite number of beautiful crystal structures. Indeed, the internet is now filled with images and videos of salt formations one can find in or around the Dead Sea: perfect cubes of halite found in the waters around the shore; salt crystals snowing down from the surface seen in underwater videos; images of salt crystals formed in odd mushroom shapes or stunning three-dimensional star-like patterns. The structures can be small enough to hold in the palm of your hand, or many feet in scale.

And it isn't just the Dead Sea where the oddities of crystalized salts can be seen. Along the Great Salt Lake, a rare mineral called mirabilite was recently observed crystalized on its shores, displaying elongated sharp crystals, looking like something between a snowflake and shattered glass. At Mono Lake, tufa towers are observed, some over 30 feet high. According to the website maintained by the Mono Lake Committee, an organization dedicated to protecting the lake, tufa towers form at calcium-rich springs beneath the lake, which interact with the carbonate-rich lake water. The result is the formation of calcium carbonate, basically limestone. Because the limestone is only formed where the springs interact with the lake water, the structures grow as vertical towers beneath the water surface. Today these towers, sometimes referred to as devils' horns, rise high above the water, easily visible, because the water surface dropped as a consequence of diversions to Los Angeles of freshwater that once flowed into Mono. It is a strange and sobering sight.

These fascinating structures have a darker side. Though stunning and arguably natural, the crystalized salt and mineral formations found near and in the Dead Sea, the Great Salt Lake, Mono Lake, and many others require a lake that is hypersaline—a lake that is either saturated in one or more salts or close to saturation. This condition is almost always human-caused; it rarely occurs without diversion of freshwater from the rivers and streams that feed salt lakes toward other uses. Such diversions lower the lake levels, increase salinities, and enable all sorts of crystallization. The problem is that these diversions, once started, are difficult to stop. Irrigators dependent on the diverted water will resist efforts to send back to the lake the freshwater they've grown dependent upon. Cities who use diversions for municipal water will likewise oppose efforts to return a saline lake basin to its original state.

One can understand the perspective of those pushing to divert freshwater from salt lakes. It is difficult to be in an arid region and watch a freshwater stream flow by. The water flowing into the salty lake can never be used to drink, to water livestock, or to irrigate and make a green field. Our planet is mostly covered with salt water in the form of oceans—why preserve these salt lakes? The problem is that when diversions reduce the flow of water into these salt lakes, evaporation does not stop. So, with continued evaporation and reduced freshwater inflow, a decline in lake level begins. This, in turn, exposes larger and larger areas of the lake bed, creating vast regions that are very salty and very dry. Due to the enormous salt content, vegetation is unlikely to grow on these flats, and dust storms inevitably result, kicking up various salts and minerals which negatively impact the health of humans and wildlife, even far from the lake location. Such problems have arisen at Mono Lake, at the Salton Sea, at Owens Lake, and at many other hypersaline lakes. The Aral Sea, however, is the most well-known of these desiccations, and appropriately so, for it resulted in a true natural disaster, one we will revisit in the chapter "Death by Human."

• • •

Though not strictly a salt lake, or even a lake at all, we would be remiss if our discussion on salt lakes neglected something called a brine pool, which could be described as a salt lake found beneath the ocean. Unlike ordinary lakes where a basin of water resides beneath the air, brine pools are located at the bottom of the sea—literally, a lake beneath the ocean. Brine pools are expanses of hypersaline water, so much saltier than the ocean that they exist as essentially isolated basins, barely mixing with the overlaying seawater. These lakes are found in regions of the ocean floor where thick layers of sedimentary salt have long been exposed. Where there is a natural depression, this salt continuously dissolves into the water, forming an extremely dense pool lying under the lighter seawater.

The odd thing about these oceanic brine pools is that the interface between the hypersaline water of the brine pool and the lighter, less saline water of the ocean above, behaves in many ways like an air/water interface—like the surface of an ordinary lake where water lies below, and air lies above. The pools can have salinities as much as eight times larger than ordinary seawater, making the water density of the brine about 1.2 times larger than the ocean. The density of an ordinary lake is roughly a thousand times larger than the air that lies above it, and so the 1.2-to-1 brine/seawater density ratio may seem small in comparison. However, the difference is sufficient to enable many of the interfacial phenomena that occur at an air/water interface. For example, deep-sea submersible vehicles can *float* on this interface just as a boat floats on an ordinary lake. When the brine level rises, brine waterfalls can be observed as the heavier brine water cascades over the rim that encloses it. Waves can propagate over the brine/seawater interface and light will reflect off of it. You can see all of these phenomena on videos available on the internet, obtained from deep-sea submersibles.* So much do brine pools resemble regular lakes that, when viewing a brine-pool video, one needs to constantly remind oneself that what resides above the brine is water, not air.

---

* As of this writing, a collection of such videos can be found at the Amusing Planet web site (www.amusingplanet.com).

A brine pool can only exist because of the density difference between two layers of water having different salinity. In this case, it is brine below and seawater above. But it isn't just salinity that affects the density of water. Temperature, salinity, and even suspended sediment can change the density of water, profoundly impacting large-scale currents in bodies of water ranging from the smallest of lakes to the entire ocean. In fact, the very life of a lake may depend on the flows induced by density differences. This is the subject of our next chapter, which also begins Part II of this book, "Life." Here we leave behind the subject of lake formation and focus on lake existence—on the processes occurring during the everyday life of a lake.

LIFE

8. LAKE OVERTURNING

Lakes have a problem. Oxygen is something that lakes need very badly, but other than that formed by photosynthetic organisms in the lake, oxygen enters a lake primarily at its surface. This would be fine if everything that consumed oxygen were also located only near the surface. But in point of fact, oxygen-consuming organisms are located throughout the water column, from the air/water interface, all the way down to the cold, dark depths, where many species of fish, among other creatures, like to spend their time.

It is an existential problem for lakes that the source of this incredibly important dissolved gas is located at a single Euclidean plane, while oxygen-breathing life lives throughout the lake volume. It is also an existential problem for the fish in your fish tank, but happily for them, you address this problem by using a bubbler, a device that pumps air from the surface down to the bottom of your tank. These bubbles then rise upward, transferring oxygen to the water as they go. But lakes don't have bubblers (although there are a few interesting exceptions to this rule that we'll look at later in this chapter). So, how is it that there is so much life in lakes, and how is it that it can be found in the depths as well as at the surface? What enables the oxygen that dissolves into the water at the lake surface to migrate downward? To understand the answer to this question, we first need to learn a bit about two physical processes, one called diffusion, and the other called convection.

Diffusion is the process whereby fluid moves from one place to another simply due to the fact that molecules are constantly bumping into each other. This can occur in a liquid or in a gas (the two referred to generically here as fluids). It can even happen in a solid, though we won't get into that here. If a bottle of perfume is opened in a very still room, a room with absolutely no air motion at all, you will soon pick up the scent at the other side of the room simply because the perfume molecules bump into each other and migrate outward from the source of perfume where they are concentrated (the bottle) toward more distant regions where they are sparse. The molecules will travel outward in all directions, eventually reaching you.

Convection is a larger-scale motion in the sense that, while it moves molecules, it does not depend on molecular-scale motion. Convection is what happens when, for example, a perfume bottle sits in front of a fan blowing the perfume-containing air directly at you. It is a much more effective, much faster means for transporting fluids in virtually any situation. Another way to understand the difference between diffusion and convection would be to take a small drop of cream and carefully place it into a cup of coffee. Diffusion will cause the white cream to gradually spread throughout the black coffee, eventually causing the two to completely mix. An example of convection would be to use a spoon to stir up the same drop of cream, a process we all know from experience is much more effective, creating a tan-colored mixture in one or two stirs.

Convective motions are often characterized as either forced or natural. The example of the fan and the spoon, above, is an example of forced convection; other examples would include any situation where a pump, fan, stirrer, or blower is used to push a fluid about. Natural convection is fluid motion that is not caused by a man-made device and occurs due to a density difference of some kind. One example of natural convection is the air movement that occurs when a cigarette burns in a very still room. Though there is no mechanically induced air motion like the example of the fan next to the perfume bottle, still there is convection. This is because the tip of the cigarette is hot, making the air immediately adjacent to the ember also hot. Hot air is lighter than cooler air and so the hot air near the cigarette tip rises upward (conveniently visualized by the smoke particles), forming what is called a *buoyant plume*. Because the rising air must be replaced, the air in the surrounding region travels inward, and a recirculating flow pattern is formed in the general vicinity of the cigarette. This natural convection is the means by which you will quickly smell cigarette smoke at the other side of the room.

In a lake of any appreciable size, the amount of oxygen transported by diffusion from the air into the water and then downward into the water depths will be quite small. As noted above, this means of transport is not very effective. Lacking a very large spoon, natural convection is needed if any appreciable amount of oxygen is to make it into the depths of a lake. But, for natural convection to occur, warmer, lighter water would need to reside on the bottom of a lake and cooler denser fluid above. If that were to occur, the heavier surface water would fall as the lighter warmer water rises. The lake would mix, and oxygen that entered at the water surface would be transported downward in a very effective fashion.

But, during the summer, the exact opposite is true. The sun heats the surface of a lake, its warming, infrared radiation unable to penetrate far beneath the surface. This creates the opposite situation from the one just described, namely a warmer, lighter layer of water above a cooler, heavier one. Such a situation is called stable stratification and though the word *stable* may sound comforting, if stable stratification is maintained, it is the death knell for life in a lake. Under stable stratification, there is still mixing in the lake. The upper layer, called the epilimnion, is mixed up by wind and wave action, and so it is a layer that is warmer, well-oxygenated, and generally of close-to-uniform temperature. But the epilimnion comprises only the upper portion of the lake, only the region where sunlight and wind have an impact.

You may have experienced the existence of the epilimnion when swimming in a lake during the summer. Away from the shore, if you dive any distance downward, you may experience a sudden temperature drop as you travel from the epilimnion into the cooler hypolimnion. The transition is often sharp, a sometimes-shocking drop in temperature. This convincingly highlights the strength of stable stratification—that such a chilly temperature can exist immediately beneath a comfortable one shows how stable stratification serves to keep these two different layers of water separate.

In addition to creating a uniform temperature within the epilimnion, mixing due to wind and waves will also create a more or less uniform concentration of oxygen in the epilimnion. Consequently, organisms that live in the upper water area are fine, assuming the temperature doesn't get too high. But those in the depths, in the hypolimnion, are living in a reservoir of cool dark water whose oxygen content decreases by the day. The separation of the epilimnion and hypolimnion can be so profound that limnologists sometimes describe a lake in summer as two lakes: an upper warmer lake that is well-mixed and well-oxygenated, and a lower colder lake that is motionless and declining in oxygen content. The two lakes live one on top of the other, in intimate contact but with little interaction.

This situation continues all summer long, solar radiation maintaining a warm light layer above a cool dark one. In the dark hypolimnion, the oxygen content slowly decreases as fish and other organisms consume it. Some of these organisms are bacteria and other microscopic organisms that break down organic matter that drifts to the bottom of the lake. Most lakes have a layer of such organic matter, sediments, typically composed of fecal pellets, leaf litter, and the bodies of any and all organisms that reside anywhere in the lake. The number of these bacteria can vary greatly, with very significant consequences to the oxygen level in the hypolimnion. Should there be a sudden spike in the amount of organic material and nutrients in the lake via, for example, a sewage spill or runoff of fertilizer or manure from agricultural activities, the bacterial population will rapidly ramp up and consume this material, also consuming large amounts of oxygen in the process. This can drop the oxygen level to a point where some organisms can no longer survive resulting in, for example, fish kills. Lacking such rapid declines, however, the oxygen level will drop in a more gradual way. Each day, the amount of oxygen decreases in a slow but continuous fashion. If this situation were maintained, all life in the hypolimnion would eventually be extinguished. But, as summer bids farewell and autumn approaches, a process occurs that is the saving grace of the hypolimnion.

Compared to the water in a lake, the air above it changes temperature relatively rapidly as the seasons change. A point is reached in the fall when a lake feels warm compared to the air above it. Once this point is reached, the air begins to suck heat from the lake surface, cooling it and reducing the surface temperature. Throughout late summer and early fall, this cooling continues, and the temperature of the epilimnion drops steadily day by day. In the depths, the hypolimnion, though cool, has warmed somewhat during the course of the summer. And so, as the epilimnion continues to cool, a point is eventually reached where the temperature of the epilimnion equals or is slightly cooler than the hypolimnion. Now the lake is unstably stratified with cold, dense, water above relatively warmer, lighter water.

This is a critical point in the life of a lake, this moment of instability. Recall the analogy of the warm cigarette ember described earlier. However, instead of a single point source of warmth at the cigarette tip, the (relative) warmth of the unstably stratified lake is distributed along the entire lake bottom. And so, perhaps on a windy day, overturning begins and mixes the lake—the surface water falls to the bottom and the bottom waters rise to the surface. And as this occurs, oxygen-deprived water from the depths reaches the surface, absorbing oxygen from the air above. Meanwhile the surface water, already saturated with oxygen, falls into the depths, providing an enormous gasp of oxygen for those creatures living in the hypolimnion.

Overturning is a very significant event in the life of a lake. It is as if this organism, this lake, gets an enormous lungful of air, a year's supply of oxygen, all in one great gulp. Some lakes overturn only in the autumn (autumnal overturning) and some overturn twice, once in the autumn and once in the spring, referred to as monomictic and dimictic lakes, respectively. Either way these one or two yearly overturns are essentially the only time the hypolimnion of a lake gets oxygen in an efficient way. You take many breaths each day. A lake takes at most two per year.

Without overturning, lakes would have a much smaller capacity for life. Oxygen-breathing organisms would live only near the surface, dramatically reducing the species diversity and the overall number of organisms in lakes. The lake depths would be relatively lifeless, dominated by anaerobic bacteria. Such situations can occur. For example, in northern climates where salt is used to prevent ice formation on wintry roads, that salt can inhibit overturning when it flows into the lake bottom. The salty bottom water becomes very heavy, so much so that even in winter, the surface layer cannot cool enough to become denser than the hypolimnion and the lake does not overturn, or only does so during periods of extremely high winds. Such a situation makes the hypolimnion very inhospitable since it not only lacks oxygen, but also contains significant quantities of salt, toxic to the freshwater lifeforms natural to the lake.

Overturning is a big deal, and authors of limnology textbooks do not fail to underscore its importance. In his 1952 textbook, *Limnology*, Paul S. Welch writes of overturning:

> *In the deeper lakes, a seasonal, thermal phenomenon occurs which is so profound and so far-reaching in its influence that it forms, directly and indirectly, the substructure upon which the whole biological framework rests...*

And all of this, overturning, stable stratification during the summer, the suppression of overturning by salt—all of it is due to density differences, the way in which the density of water is affected by temperature and salt.

● ● ●

The spring overturn is a bit different from the autumnal one, and not all lakes have one. To have a spring overturn, the lake must become stably stratified once again during the winter, just as it does during the summer. But this happens in a much different way. In climates where temperatures fall far enough for a lake to freeze over, the lake will spend most of its winter sequestered from the atmosphere. The upper boundary of the liquid lake is the ice/water interface, and right at that ice/water interface, the temperature will be 32 degrees F. If the air temperature above the ice drops below freezing, as is often the case, that air will suck more heat from the lake, and the ice sheet will get thicker by transforming liquid at the ice/liquid interface into more ice. But, the temperature right at the ice/water interface will not deviate significantly from 32 degrees F. This is important because at 32 degrees F., water is lighter than its slightly warmer counterpart since the maximum density of water exists at 39 degrees F. So, prior to freezing over, as a lake cools, the point at which a lake reaches 39 degrees F., is the point where it stops mixing. When the lake is above 39 degrees F., cooling at the surface results in the formation of denser water, which falls to the bottom of the lake, replaced by warmer lighter water. This continues until the entire lake achieves a temperature of 39 degrees F., after which mixing stops.

Once the lake reaches this point, any further cooling makes the lake thermally stratified. Further cooling does not cause the cooled water to fall downward, since this water is lighter than the 39 degrees F. water. Hence, further removal of heat at the surface only cools the very surface since the cooled fluid is now lighter than the water beneath it. Eventually the surface hits freezing, an ice skin forms, and the lake typically stays in this state with 39 degrees F. water at its bottom and 32 degrees F. water at the ice/water interface, with a cooling transition as you travel from the lake bottom to the top.

As spring approaches, the lake will still be stably stratified. The bottom will be dense, at 39 degrees F., and the top somewhat lighter, at 32 degrees F. In the spring, as the ice melts off the lake surface, a situation occurs that is similar to summer's end. Whatever oxygen-breathing life that has survived the winter has done so by relying on oxygen remaining in the lake when it iced over. As the ice melts away, the stratification still exists. As the sun begins to warm the surface of the lake, it increases the surface water temperature from freezing. As the surface water gets warmer, it does get heavier, but until it hits 39 degrees F., it remains lighter than the water in the hypolimnion. However, this heated surface water *is* heavier than the water just beneath, and so the upper layer of the lake will mix with that just below it.

This process will continue, the surface waters getting progressively warmer, and that warmth penetrating progressively downward, mixing with the water just beneath. At some point, the entire lake will be at 39 degrees F., a point at which the lake will have no density gradient at all. The temperature will be everywhere the same, and the density will be everywhere the same. Under these conditions, much like for the autumnal overturn, any bit of wind will cause the lake to mix since there is no stratification to resist that mixing. And, since it is often windy in the spring, and this wind can be robust, once overturning occurs the lake will tend to mix thoroughly, oxygenated water at the surface being transported downward.

● ● ●

The above description of overturning is necessarily simplified. Though many lakes follow the process described above to a great degree, there is also significant deviation. For example, during the early portion of summer, a moderately stratified lake might become mixed should a violent storm pass over. Streams, waterfalls, and rivers that run into lakes may provide a source of cool, oxygenated water to the hypolimnion, even during stratification. Weather patterns and precipitation and wind affect the degree of stratification as well as the vigor of overturning. Lakes may have a spring overturn one year, but not the next. Lakes may only partially freeze in the winter, or not at all. Large lakes may exhibit significant variations in their depth: parts may overturn while other parts do not. Some lakes are too shallow to have overturning at all. Some lakes are too deep for the hypolimnion to ever participate in overturning. A whole host of permutations of stratification and mixing exist, which has resulted in the evolution of a wonderfully rich vocabulary to describe processes associated with lake overturning. For example, following the description provided by Gerald A. Cole in his 1994 *Textbook of Limnology* there are these categories for lakes and lake processes associated with overturning:

- **Amixis:** Amictic lakes are lakes that never mix. Examples of such lakes include those continually covered in ice and therefore are never exposed to wind, preventing mixing.

- **Holomixis:** Holomictic lakes undergo overturning at some point during the year and where, importantly, the entire lake is involved in the mixing process. This broad category covers many of the world's lakes.

- **Oligomixis:** In the tropics and at low altitudes, there is little variation in temperature, the average daily temperature being relatively warm and constant throughout the year. Lakes in this category tend to be warm above and cool below and there is no cold season to enable destratification. Oligomictic lakes rarely overturn and mixing in them is unusual. Very deep lakes can also be oligomictic, such as Lake Tahoe. These lakes may experience large seasonal variations in temperature, but due to their depth, they still may not undergo complete overturning.

- **Monomixis:** Monomictic lakes have one overturn per year. This could be either a spring overturn, or an autumnal overturn. The former is called cold monomixis, and the latter warm monomixis.

- **Dimixis:** Dimictic lakes have both a spring and autumnal overturn.

- **Polymixis:** Polymictic lakes undergo mixing many times during the year or may even overturn continuously throughout the year. Typically, polymixis occurs when the day to night temperature difference is large enough to cause stable stratification during the day and then significant surface cooling at night, which makes the water column unstable and causes overturn. In such lakes, this day-to-night variation in temperature may be larger than the summer-to-winter temperature difference. Such lakes often occur at very high altitudes or in desert climates.

- **Meromixis:** Meromictic lakes undergo overturning, but the entire depth of the lake does not participate in the overturn. An example of this is the Dead Sea, further described below.

Of all these, this last type of lake, the meromictic lake, is perhaps the most interesting, though not the most common. Such lakes typically have a deep layer called the monimolimnion where some dissolved substance, often salt, keeps the water very dense and prevents any cooling of the lake surface from increasing surface-water density to a value equal or greater than that in the monimolimnion. The water above the monimolimnion may undergo mixing, but not with the monimolimnion, at least as long as meromixis holds. This region is typically devoid of oxygen and hosts only anaerobic life. So, the monimolimnion is not very lively; it is an isolated, dark place where the atmosphere above may have no impact for centuries, or even millennia.

An interesting example of a meromictic lake, at least until recently, was the Dead Sea, which we discussed in the last chapter. For at least 300 years prior to 1979, the Dead Sea was meromictic, with a very salty, cold body of water in its depths. Cooling of the water surface was never sufficient to bring the surface-water density to a value higher than that of the depths. Accordingly, the dark, salty bottom of the Sea stayed unexposed to any surface influence for centuries. As noted in the previous chapter, freshwater diversions from the streams and rivers that feed the Dead Sea served to reduce the amount of freshwater flowing in. This freshwater, being lighter than the salt water it was flowing into, tended to stay near the surface and helped keep the density of that surface water low. Hence, freshwater inflows maintained the Sea's stable stratification by keeping the surface lighter than the depths. However, the diversions of this freshwater caused the lake surface to become saltier, its salinity slowly approaching that of the hypolimnion. This process occurred throughout the 1960s and 1970s, and in 1979 the density at the surface and in the hypolimnion equalized.

In February, the lake overturned. In the centuries prior to this, the bottom of the Dead Sea had been essentially devoid of oxygen, the only life being anaerobic in nature. Such life forms excrete hydrogen sulfide (also known as rotten egg gas). This gas was released, and not just a little hydrogen sulfide—300 years' worth of hydrogen sulfide. The stench must have been awful, but at least it helped confirm why a 10th-century geographer had named this lake the Stinking Sea.

• • •

Generally speaking, meromixis implies some sort of ill health for a lake—something not very useful. While this is probably more or less true, the anaerobic bacteria that thrive in the dark, cold, often salty depths of these lakes would disagree. So perhaps we would be remiss if we left this discussion of mixing without mentioning something called a solar pond, a typically manmade body of water, deliberately constructed to create meromictic conditions, but for positive purposes.

Solar-energy projects are often stymied by the difficulty of collecting the energy that the sun provides us so prodigiously. On average, slightly more than one kilowatt of energy strikes every square meter of much of the planet, but how to collect it? Solar panels have, in recent years, seemed to take over this task, converting the energy directly into electricity. But in the past, the focus was on an alternative approach to this problem (and there are some current efforts in this regard as well), namely, to use solar energy to create a superheated vapor. This is the typical way in which electricity is produced, but via the combustion of fossil fuel. In this more typical situation, the energy released from combustion is used to boil water and create superheated steam, which drives a turbine that turns a generator, thereby making the electricity that is funneled into the power grid. The overwhelming majority of electricity you have ever consumed was created in this fashion.

An interesting question one can pose is whether solar energy could be used to boil water or, if not boil it, at least heat it to the point where it could be used to do something useful. Anyone who has ever set foot in a kiddie pool left sitting in the sun for the better part of the summer would probably intuit that this is unlikely to occur. No matter how long you keep a kiddie pool outside, its water rarely gets terribly hot, let alone boils. And the reason is that the sun penetrates to the bottom of the pool, heats the floor, makes the water adjacent to the floor warm and light, causing it to mix. The bottom water rises upward, mixing the pool. In other words, a pool or shallow pond exposed to the sun is constantly warmed and mixed, and the warm water at the surface evaporates, serving to cool the pool and cause the water level to fall. Nowhere does the fluid get truly hot.

There is a very simple way to counteract the loss of heat from a shallow pool and that is to place a large amount of salt along its bottom. This creates a situation much like the Dead Sea pre-1979, but with the important difference that solar ponds are shallow enough to allow sunlight to penetrate to the bottom. Because of this, the sun does heat the bottom of the pool. But as the bottom water gets hot, it also dissolves large quantities of salt, making it heavy. Though hot, the salt makes the water heavier than the cooler water above. In this way, the bottom of the solar pond keeps getting hotter and saltier and heavier and therefore does not rise to the surface. The fresh water at the surface of the pool remains cool and relatively light. Being cool, the surface also evaporates more slowly. This is stable stratification and it enables water at the bottom to get very hot, approaching but not quite achieving the boiling point of water.

Commercial power plants boil water at high pressure, transferring tremendous amounts of energy into the superheated steam, which is then used to drive a turbine. Doing this with a solar pond is not possible. Although temperatures higher than those obtained without using salt are attained, they are not hot enough to boil water and create superheated steam. To deal with this, an approach can be taken where the heat from the hot salty water is transferred to a liquid that has a low boiling point, which boils and generates vapor that can drive a turbine. Attempts have been made to generate electricity in this way, but they have not caught on. Still, it is a fascinating example of how density gradients, and the manipulation of density gradients, can change the way in which water does or does not move about.

• • •

Earlier in this chapter, we noted how maintaining oxygen levels in a fish tank or aquarium is often achieved via bubblers, something which lakes do not, in general, have. But, as noted earlier, there are exceptions to this rule. In some situations, large-scale bubblers have been installed in lakes, as a means to do what nature will not.

For all but a short portion of its length, the border between South Carolina and Georgia is defined by the Savannah River. Like many rivers, the Savannah has been dammed at several locations to create lakes, artificial reservoirs used for flood control, hydropower, and recreation. These are Lake Hartwell, Richard B. Russell Lake, and J. Strom Thurmond Lake. Like most public works projects of this size, the building of a dam and the flooding of thousands of acres of land involved eminent domain, no small amount of conflict, and quite a bit of horse trading among numerous parties. One concern regarding Lake Russell involved the trout fishery downstream of the dam. Sustaining these fish would require dissolved oxygen concentrations of 6.0 milligrams per liter or greater, and a temperature no greater than 70 degrees F. Such temperatures are best found in the hypolimnion, and so releasing water through the dam, from the hypolimnion of Richard B. Russel (RBR) Lake into the region downstream of the dam was most likely to realize those cool temperatures. But, for a significant portion of the year, the oxygen concentration in the hypolimnion was expected to be less than 6.0 milligrams per liter, thereby missing the required conditions. To address this, the U.S. Army Corps of Engineers designed and implemented a system to increase the oxygen content in the hypolimnion of RBR Lake so water released from the dam would have the requisite temperature and oxygen concentration to enable trout survival.

The system built at RBR Lake consists of an injection system on the upstream face of the dam that enables short-term addition of oxygen when needed, along with a more continuous injection system located at the bottom of the lake and about a mile upstream from the dam. Pipes run from large oxygen-storage canisters located on the Georgia side of the lake into the lake bottom where they terminate in large bubblers designed to create clouds of small bubbles. The goal of these two systems is to ensure water leaves the dam with a minimum of 6.0 milligrams per liter of dissolved oxygen. The system is typically used in the summer when the lake is stratified and oxygen levels in the hypolimnion are low. In a 1994 report describing the water quality of RBR Lake, the U.S. Army Corps of Engineers noted that in one year (1988), the oxygenation system was turned on May 11 and turned off on November 9. During that period, oxygen was delivered to the lake at rates from 15 tons of oxygen per day to as high as 65 tons per day. Indeed, the peak capability of the system is quoted at 100 tons of oxygen per day!

While the oxygenation system installed at RBR Lake may seem odd, such installations are not uncommon. The overall problem solved by these systems concerns the fact that, generally speaking, a free running river is a well-oxygenated river. When reservoirs are built by damming up a river, the rapid flow, turbulence, and bubbles characteristic of a flowing river cease, as does the associated oxygenation. Hydropower projects therefore necessarily require issues of oxygenation be dealt with, or at least considered. Nor is this a new problem. Indeed, the first reported use of this approach, often referred to as hypolimnetic aeration, occurred in 1949 at Lake Bret in Switzerland. It is a way humans seek to control oxygen concentration, one of the dissolved gasses critical to the overall ecological health of a lake. But oxygen can rarely be thought of by itself when looking at the ecosystem of a lake for, tied closely to it, is carbon dioxide. As we will see in the next chapter, the delicate interplay of these gases is critical to the life of a lake.

9. DISSOLVED GAS

Oxygen and Carbon Dioxide

Look at a tumbler of water, fresh from the tap. What's in there? Mostly water, certainly, but also present, and just as colorless and invisible are the dissolved gasses. A lot of them are there, both in the water in your glass—and in the water in any lake you will ever see. Dissolved in this water you will find nitrogen, oxygen, carbon dioxide, argon, neon, hydrogen sulfide, methane, and others. The range in concentrations of these gases is large, and their impact on the chemistry and biology of lakes varies from almost no impact at all to critically determining the ability of water to support life. Just which are present, in what amounts, and in what ratios—all of this is part of the secret, transparent life of dissolved gases.

But oxygen and carbon dioxide are the real players. These two gasses and the processes involved in their creation, consumption, and transport can explain a huge amount of the biological happenings in any lake. On the one hand you have photosynthesis, which consumes carbon dioxide and creates oxygen. On the other hand, you have respiration, the process used by aerobic organisms to survive wherein oxygen is consumed and carbon dioxide is created. Given that every living organism in a lake is involved in one or both of these processes (with the exception of chemosynthetic organisms, which we will ignore here), oxygen and carbon dioxide can be thought of as the currency of the limnological world.

In some ways these two critical gasses can be thought of as mirror images of each other, two entities that exist in a kind of yin and yang relationship that mediates, and is mediated by, life. Via photosynthesis, plants consume carbon dioxide and create oxygen, while via respiration, other organisms (including plants) consume oxygen and create carbon dioxide. But this view of carbon dioxide and oxygen as mirror images of each other is true in only the most general sense because these gasses differ in several ways. Though both gases are critical to life in lakes and control to a large degree how and why lake life is the way it is, they behave differently. For one thing, carbon dioxide actually reacts with water, while oxygen does not. Once dissolved in water, oxygen is just oxygen in solution, it does not react with the water molecules. But once dissolved, carbon dioxide reacts to form carbonic acid as well as bicarbonate, and this has wide-ranging consequences. The addition of carbon dioxide to water results in significant deviations of the water pH from neutral (pH = 7) and, combined with elements such as calcium and magnesium, can result in complex buffering solutions that impact the amount and type of life that can exist in a lake. Depending on other conditions, this chemical system can result in the dissolution or deposition of carbonates, such as calcium carbonate (limestone).

The solubilities of oxygen and carbon dioxide are also quite different, and their concentrations in air are different, both of which impact the amount of these two gases that will be found in water. Air is 21 percent oxygen and only 0.04 percent carbon dioxide, a factor of 525 difference. Indeed, notwithstanding society's concerns about the growing level of carbon dioxide in the atmosphere and its impact on our climate, it should be noted that there is so little carbon dioxide in air that the number of argon molecules in a typical air sample exceeds the number of carbon dioxide molecules. But carbon dioxide is much more soluble in water than oxygen, and so a sample of water saturated with air will contain 1.3 milligrams per liter of carbon dioxide and 14.2 milligrams per liter of oxygen (at sea level and freezing), only a factor of 11 difference.

This is one reason why carbon dioxide is found in such abundance in water, despite its very low concentration in the air above.

Life, of course, has a huge impact on the concentrations of both oxygen and carbon dioxide. In lakes containing a lot of life (often referred to as productive lakes—more on this, below), the amount of oxygen and carbon dioxide in the water may deviate significantly from the saturation concentrations cited above due to the formation of oxygen via photosynthetic organisms, and the consumption of oxygen by aerobic organisms. Conversely, for lakes that have very little bioactivity, such as rocky, alpine lakes with very clear water, oxygen and carbon dioxide concentrations may be essentially identical to their saturation values.

The concentration of these critically important gasses will vary from the surface to the lake bottom and from one portion of a lake to another. They will also vary over the course of a day, a season, and a year. Even in a lake completely devoid of life, variations in the amount of oxygen and carbon dioxide can exist. This is due to the sensitivity of gas solubility to temperature, these solubilities increasing as temperature decreases. For example, at 86 degrees F., the concentration of oxygen and carbon dioxide in a sample of water saturated with air are 7.5 milligrams per liter and 0.5 milligrams per liter, respectively. However, at freezing these concentrations increase to 14.2 milligrams per liter and 1.3 milligrams per liter respectively, roughly a factor of two increase for both gasses. Hence, at least near the surface, the simple daily change in temperature from its peak in the afternoon to its minimum before sunrise can result in significant changes in gas concentration.

● ● ●

In many lakes, the real determinant of oxygen and carbon dioxide levels is the amount of life in that lake. And the amount of life in a lake can vary quite a bit. In the early days of limnology, scientists struggled to understand the obvious differences that could be found in lakes.

In high-mountain regions where lake bottoms were rocky, the water could be crystal clear and the amount of life present very small. In contrast, in typically lowland areas with significant runoff from surrounding slopes, where lake bottoms were filled with organic material, and where lakes were shallow, the water would be murky and filled with all kinds of plant life, fishes, and amphibians—everything from the microscopic to the largest lacustrine mammals. Various terms have been developed to describe these two types of lakes, the most common being oligotrophic and eutrophic, respectively. Although there are many lakes that cannot be clearly classified as one or the other, the general concept that a lake exhibits either eutrophy or oligotrophy continues to be an important organizing idea in limnology.

Eutrophic lakes are often shallow. They have high nutrient levels, abundant planktonic life and large concentrations of attached algae. Here the word *nutrient* is used to denote inorganic material necessary for life (such as phosphate, nitrate, and silica). These lakes may have surface blooms of blue-green algae, which can give the water a yellow-green color, and the water clarity is, in general, low. In eutrophic lakes the oxygen in the hypolimnion becomes depleted in both the winter and summer in temperate regions, but near the surface, in what is called the photic zone where enough light penetrates to allow photosynthesis to occur, the water has abundant quantities of oxygen and can even become supersaturated with oxygen during the day. Indeed, it is possible to see small bubbles near the shore of such a lake consisting of pure oxygen formed by attached algae. At night the oxygen level falls below the saturation level as organisms that consume oxygen do so without the replacement of that oxygen by photosynthesis. In these eutrophic lakes, there is an abundance of species in all parts of the food chain, including fish. Eutrophic lakes, in short, simply brim with life.

Oligotrophic lakes are essentially the opposite of eutrophic lakes. They tend to be deep with very clear water and very low nutrient levels. Because the total amount of life existing in an oligotrophic lake is small, the concentration of dissolved gases tends to be determined by the atmosphere and not the biosystem present, and so the amount of oxygen tends to be at the saturation level throughout the year as well as throughout the water column.

Before we proceed, it is important to note that neither eutrophy nor oligotrophy (nor mesotrophy, a catch-all descriptor for lakes that fall between the two) should be thought of as the "good" state of a lake. Aesthetically, oligotrophy may seem like the desired state of a lake with all of the crystal-clear water such lakes can contain. But such lakes have very little life in them, and their ability to support large numbers of fish is limited. Eutrophic lakes, on the other hand, though often muddy in appearance and with sedges, algae and various pond scums along their shores may be aesthetically unappealing but are often full of life. Neither is necessarily better than the other.

However, it is also true that human activities can shift the trophic state of a lake and this is typically harmful. Runoff of fertilizer from a nearby farm field, the introduction of sewage, and other activities will introduce nutrients to a lake. Such intrusions tend to move a lake in the direction of eutrophy, a process called eutrophication. For example, the introduction of phosphorus to an oligotrophic lake via fertilizer runoff will, at least for a period of time, increase the amount of life in a lake. Such a spike in the nutrient level in the lake may result in massive algal blooms, which, in turn, will temporarily increase the oxygen levels near the surface. This may sound like a positive thing. However, once the algae dies, the blooms sink to the bottom. There they undergo decomposition by oxygen-consuming bacteria, removing large quantities of oxygen from the hypolimnion and potentially causing a fish kill. Hence, such manmade changes in nutrient levels might move an oligotrophic lake toward eutrophy in a harmful way. However, this is not harmful because being eutrophic is inherently bad, it is harmful only because the human activities shifted the lake from its original state to a new state where the extant life forms may not be able to survive.

• • •

Broadly speaking, the aforementioned definitions of eutrophy and oligotrophy enable one to categorize lakes with relative ease. But limnologists typically want to know more than whether a lake has a "lot" of life in it, or "not very much." Rather, what is desired is to quantify precisely how much life exists in a lake. Such a quantification then enables limnologists to study and discuss how the quantity of life in a lake is affected by a range of parameters and situations such as season, geographical location, altitude, chemical environment, pollutant levels, and many other factors. In limnology and the ecological sciences, all of this is referred to as production or productivity. Specifically, productivity is understood to be the rate at which life is formed in a lake, and more formally as the net increase in biomass per unit time. Productivity can also be characterized in terms of energy (as in the energy in the chemical bonds of the organic matter created per unit time).

Productivity is an important concept in ecosystems and, as it turns out, it is also a hard thing to quantify. If productivity is taken to be the amount of biomass created per unit time in a lake, then its quantification is a challenge for several reasons. First of all, organisms in lakes can be extremely small, making their collection a challenge. Moreover, they are wet and separating biomass from water can be difficult.

A more straightforward way and one that is, for our study of dissolved gases, more illuminating, is to quantify productivity in terms of the amount of oxygen produced. Because life, be it plant life or animal life, is intimately involved in either the production or consumption of oxygen, tracking the amount of oxygen formed provides an avenue for effectively quantifying and tracking photosynthetic production. To do this, let us first look a bit more closely at photosynthesis, recognizing that all life in a lake (or any other ecosystem for that matter) can trace its biomass to photosynthesis. This is obviously true for plants. But it is also true for herbivores, and also for carnivores, since what they consume either consumed a plant or consumed something else that consumed a plant, and so on. Hence, biomolecules that make up any living thing all have their origin in photosynthesis.

Let us simplify the process of photosynthesis using the oft-cited chemical equation:

$$n\mathrm{CO_2} + n\mathrm{H_2O} + \text{light} \rightarrow (\mathrm{CH_2O})n + n\mathrm{O_2}$$

Here, the term $(CH_2O)n$ refers to any of a number of molecules that can be formed during photosynthesis. For example, if $n = 6$, the molecule formed is hexose. The important point is that photosynthesis takes this inorganic gas, carbon dioxide, and water molecules, and creates complex biological molecules, along with oxygen. It is this $(CH_2O)n$ which we call biomass in production. And, it is this simple equation that is really the miracle of all life on earth. In one form or another, every biomolecule in every organism on Earth can trace its existence to this reaction, to photosynthesis. This is one of those things in science that we learn at such an early age—typically in a junior high school biology class—that we tend to forget how absolutely incredible it is. Everything that is alive—every single biomolecule in a mouse or a protozoa or a human can trace itself back to this thing called photosynthesis. The biomolecules making up the cells that comprise one of the capillaries in the tip of your pinky came from something you ate, perhaps a steak. The steak came from a cow whose biomolecules came from grass or grain or corn. The biomolecules in those plants in turn were all formed by photosynthesis—this wonderous process that utilizes the energy contained in the photons radiating from the sun to form complex, high-energy molecules. Plants create these biomolecules out of carbon dioxide and water. In other words, every single carbon atom in your body came from a molecule of carbon dioxide, traceable to photosynthesis. It is the most miraculous of things. Without this one chemical equation, Earth would be nothing but a rocky wasteland.

A good question one might ask is, what keeps photosynthesis in check? In other words, since the sun has been shining on our planet for the billions of years of its existence, and certainly ever since photosynthetic life first appeared, what keeps the amount of biomass finite? Why is our planet not covered with miles-thick layers of biomolecules? Why aren't the oceans clogged to their depths with photosynthetically created algae and other plants—a veritable slush of green ooze from the ocean surface all the way down to its bottom? Most readers will quickly recognize that the piling up of plants will ultimately result in their decomposition, something anyone who has built a compost pile understands in an almost reflexive way.

But why do plants rot, or decompose, or break down? The answer is one word: respiration. Plants are consumed, either directly or indirectly, by other organisms that use respiration to release the energy created in the photosynthetically created biomolecules, to create biomolecules of their own. A fish may eat a bit of plant matter, break it down and via respiration release some energy, which is used to create a biomolecule that the fish needs—to create more fish. A cow uses respiration to turn grass into more cow. An unbelievably large number of bacteria exist absolutely everywhere on Earth, trillions of trillions of trillions of bacteria, busily breaking down dead and dying plants and using oxygen to transform the biomatter of the plant into energy, thereby creating bacterial biomass. The equation for respiration is:

$$(CH_2O)n + nO_2 \rightarrow nCO_2 + nH_2O + \text{energy}$$

This equation is essentially the reverse of the photosynthesis equation presented above except that, on the left, the molecules consumed are much more variable than those formed in photosynthesis. They may include any organic molecule. On the right, instead of reforming the light used to create the molecules in photosynthesis, what is now released is energy. Some of this energy, a very small fraction in fact, is used to create new biomass in the organism doing the respiration.

But life is very inefficient in its use of energy, and so most of the energy on the right side of the respiration equation is dissipated in the form of heat. If, for example, the organism we are considering is you, the energy produced as you use the oxygen you breathe to break down the food you consume will be used to rebuild and maintain your body. But that is where a very small fraction of the energy goes. Most of that energy is simply dissipated as heat. As noted by Cole in his 1994 *Textbook of Limnology*, the energy contained in the biomass of a fish is less than 0.1 percent of the energy contained in the plant matter consumed by the invertebrates that served as its diet. The overwhelming majority of the energy fixed in the biological molecules created by plants is ultimately lost, dissipated as heat to the surrounding environment and ultimately, radiated back into space, lost to the universe.

Before moving forward, we would be remiss not to talk just a little about other ways you may be familiar with the respiration reaction described above, without really knowing it. If you were to leave a piece of wood out long enough, microorganisms would infest it and consume the organic molecules in that log, leaving very little behind, the remnants of the log looking much like topsoil and comprising the dead bodies of bacteria and other microorganisms that fed on those bacteria. This process can also be referred to by another name, one not often used in biology, namely oxidation. This should make sense after a bit of thought. After all, we are taking a biomolecule and combining it with oxygen to create heat, carbon dioxide and water. This is oxidation.

Oxidation can occur without the help of life. Take, for example, the coloration of a white piece of paper you may have left on the dashboard of your car. Over time, you will observe it turn tan and eventually dark brown. This occurs because of the presence of oxygen—the paper is being oxidized. Now, if we were to raise the temperature at which this reaction takes place, we would call it by yet a different name, namely combustion, or burning. This is what happens when you put a match to a piece of paper; you are increasing the temperature of the cellulose, enabling the oxidation reaction to occur much more rapidly than it would if that same paper simply lay on your car's dashboard, nominally at room temperature. In either case, the cellulose is being oxidized to form carbon dioxide, water vapor, and heat.

The unique thing about combustion, namely the existence of a flame, concerns the fact that the reaction moves more quickly as the temperature rises. Because the reaction moves more quickly, more heat is released per unit time, which serves to raise the temperature. A critical temperature, the flame temperature, is reached when the amount of heat released maintains the fuel at this elevated temperature. If the paper continues to burn after the flame from a match is taken away, then that critical temperature has been reached. For paper, this temperature is 451 degrees F., as you may know from reading in your high school English class the short novel *Fahrenheit 451* by Ray Bradbury. When you blow on a flame, such as a candle flame, the reason it "goes out" is because the relatively low temperature of your breath has cooled the wick down to a temperature beneath that critical temperature. This drop causes the wick to quickly approach room temperature, where oxidation continues though at a dramatically slower rate.

So, after some thought, you will realize all of these things are related: the decomposition of organic material by bacteria; the yellowing of paper on your car dashboard; the combustion of wood—they are all the same chemical reaction by different names: respiration, oxidation, combustion, burning, etc. An important difference between them is the rate at which the process occurs. It will take years for a log to decompose in a forest, but only minutes to burn in a campfire. Another important difference: combustion doesn't create anything. When the paper has finished burning, all that is left is the heat that was released, carbon dioxide, water vapor, and some ash. When respiration occurs in your body, carbon dioxide and water are formed, and some heat is released, but also more of *you* is created.

Now, let us get back to lakes.

To understand the productivity of lakes, lake ecologists spend a lot of time looking at photosynthesis and respiration. The rate of photosynthesis and respiration are given the variable names *P* and *R*, respectively. These values are often estimated as the amounts of oxygen created and consumed, respectively, and which can be measured with reasonable accuracy via a variety of techniques. The ratio of the two, the *P*/*R* ratio, is a parameter used to characterize lakes. Productivity is closely related to *P*, and so *P* is used as a proxy for it, since the amount of oxygen created via photosynthesis is proportional to the amount of photosynthetically produced biomass, the ultimate source of all biomass.

One might presume that *P*/*R* should always equal one, since a value greater than one would result in an increase in oxygen concentration without bound, and a value less than one would result in a value of oxygen concentration that would drop to zero. And, if one considered a large enough ecosystem and averaged over a long enough period of time, a value of $P/R = 1$ or close to it would likely result.

But for actual lakes, which are necessarily of limited size, *P*/*R* can deviate from one. Moreover, *P*/*R* can vary with time of day, season, and other conditions. These deviations from $P/R = 1$ are due to a variety of reasons. For example, let us imagine a lake located in a forest with trees overhanging its shores. Enormous quantities of leaf litter will fall into the lake and ultimately settle onto the lake floor. Large quantities of leaves and similar organic matter will also wash into the lake during heavy rain events. These leaves, these pieces of plants, were not formed in the lake. But they will be broken down by the lake's bacteria, which will consume oxygen, causing *R* to be large, though *P* may be relatively small, at least for some period of time. Of course, if one considered *P* to include the surrounding forest, its value would be much bigger, causing *P*/*R* to get closer to one, but this is not the way things are done in limnology.

One might worry that a lake such as the one we've described, having a leaf-litter covered floor, would eventually shut down. This is a logical expectation given that the amount of oxygen present in a lake is finite and if photosynthesis is less than respiration, then the oxygen level would have to fall to zero. This would indeed be the case if the only way a lake receives oxygen is via photosynthesis. But, it is possible for oxygen to enter a lake in many other ways, via, for example, the flow of well-oxygenated streams into the lake, from runoff on its banks, and from transport of air across the air/water interface.

But, still, it may be the case that a lake does shut down. For example, leaf litter that falls to the lake floor may be quickly decomposed by bacteria, at a rate that consumes oxygen faster than it enters the lake via any of the methods described above. This could cause the bottom of the lake to become anoxic, lacking in oxygen, effectively shutting down. This, however, may not be as devastating as it sounds. At the surface, photosynthesis may be happily occurring, and lifeforms relying on the oxygen it creates near the surface may do just fine. Furthermore, at some later time, once oxygen has been returned to the depths, decomposition of the leaf detritus can resume.

Lakes can also have a value of P/R greater than 1.0. This occurs when there is plenty of sunlight and the amount of plant biomass production exceeds what is consumed by plant-eating organisms. One might expect this situation would ultimately result in a lake completely clogged with plant matter, and in a certain very long-term sense this can be true. The plant matter that is not consumed typically falls into the sediment at the bottom of the lake where some may be consumed by aerobic bacteria. But, some of the biomatter may simply accumulate, slowly creating an ever-thicker layer on the lake floor, perhaps to be partially consumed at some later time when a high oxygen level occurs in the hypolimnion. Or perhaps, much of this layer will simply cause the lake to fill in, though it should be noted that the rate at which such sedimentation occurs is typically on the order of millimeters per year. But this is a pathway by which a lake can die, something called senescence, which we'll return to in later chapters.

Oligotrophic lakes generally have a *P*/*R* ratio close to one. Both *P* and *R* will be relatively low for these lakes. The photosynthetically generated biomass is consumed by a relatively small density of organisms, including, typically, a small fish population. The water stays relatively clear, and the bottom of the lake is free of significant quantities of sediment. Eutrophic lakes, on the other hand, tend to have *P*/*R* ratios greater than one. Photosynthesis in the upper layers of the water produces abundant biomass. However fecal pellets produced by organisms that consume this and dying plant matter, all sink to the bottom where there may be insufficient oxygen to degrade them, again, resulting in a gradually thickening sediment floor. There is some irony here; a lake that is so photosynthetically active and therefore forming so much oxygen does not have enough oxygen to consume the uneaten plant matter at its floor. This situation simply highlights the fact that lakes are not well-mixed test tubes, and the oxygen present at the surface is obviously not available to organisms in its depths. Lake overturning, described in the previous chapter, is therefore critically important.

Of course the *P*/*R* ratio is only so useful. Two lakes may have the exact same value for *P*/*R*, though in one lake both *P* and *R* could be ten times larger than in another lake. And so, lakes are sometimes categorized on a plot of *R* versus *P* with a line having a slope of one (i.e., where $P = R$ and hence $P/R = 1$) bisecting the diagram. Lakes having a value for *P*/*R* greater than 1.0 or *P*/*R* less than 1.0 will appear as data points located below and above that line, respectively. Much more productive lakes will appear in the upper right side of the diagram, while less-productive ones will appear on the lower left side. For example, a lake that receives partially treated sewage and thus is suffused with nutrients, but which has adequate aerobic life to consume the algae that will inevitably bloom in the presence of such nutrients may have a value of *P*/*R* close to one; it will appear in the upper right quadrant of the diagram. Meanwhile, an oligotrophic lake may also have *P*/*R* close to one, but because it has few nutrients in its water and therefore has very low rates of production and consumption of oxygen, appears in the lower left portion of the diagram.

● ● ●

Lakes are largely stationary bodies of water lacking in significant turbulence, waves, and whitewater, at least for all but the very largest of lakes. This is worth noting, because it explains why some of the complexity we have discussed in this chapter exists. If a lake was somehow well-stirred, that is to say its surface was always being chopped up via waves and wind, and the resulting bubbles were being entrained and transported throughout the lake, all the way to its very bottom, the dissolved gas concentrations in the water would always be at or very near the saturation values, from the lake surface all the way to its bottom. It wouldn't matter how much life existed in the lake, the amount of oxygen and carbon dioxide would be at their saturation levels. In this contrived situation, the lake would always have enough oxygen for its aerobic community to consume whatever organic matter might come its way, independent of the amount of oxygen formed via photosynthesis. For example, the introduction of a high nutrient load from human activities such as sewage or fertilizer would be quickly consumed by aerobic bacteria that would have all the oxygen needed to break down the organic matter; they wouldn't have to wait for photosynthetic organisms to provide the oxygen necessary to complete this task.

This is a very different situation from actual lakes in that there would be a decoupling of *R* from *P*. And though this situation does not occur in lakes, something akin to this does occur in streams and rapidly moving rivers, which are very effective at dissolving air into the water. The combination of bubble formation and turbulence ensures a rapid and continuous transfer of air into the water as dissolved gas. Accordingly, streams usually have oxygen concentrations close to their saturation values, which has a benefit when it comes to the tolerance of a body of water to pollution. The presence of oxygen results in the decomposition of organic constituents, and the consumption of oxygen in the process is readily replaced by more oxygen via bubble aeration, whitewater, and turbulence. Undoubtedly, this is where the notion that fast-moving water is safer to drink than stagnant water, a guide that is useful only in the most desperate of situations. Hikers note that filters or iodine tablets should always be used.

10. ICE

In 1963 Kurt Vonnegut wrote *Cat's Cradle*, a novel about a future world where a new kind of ice has been developed. He imagines "ice-nine," a form of ice stable at room temperature and capable of transforming liquid water into room-temperature ice via simple contact. Ice-nine is of course incredibly dangerous and Vonnegut details the catastrophes that inevitably ensue. Happily for us, the existence of an ice that could solidify every organism, ocean, and lake remains within the realm of science fiction. Indeed, our own ordinary ice, which melts and freezes at 32 degrees F. is actually quite friendly to our biosphere, in many ways serving more as a protector of life than a threat to it. And while the idea of ice that is stable at room temperature may fascinate, we should hasten to note that the ice we see form every winter is pretty fascinating stuff in its own right.

In many ways, most of the miracles one finds in lakes can be traced back to one single miracle, which is the water molecule. It is a simple enough thing, just one oxygen atom bound to two hydrogen atoms. But the three are not bound together in a straight line; rather, they are oriented at an angle of 104.5 degrees. Think of the water molecule as your arm, the hydrogen atoms are your fist and shoulder, and the oxygen atom is the elbow. If you hold your arm straight out, the hydrogen atoms are oriented at an angle of 180 degrees from each other. If this were the actual orientation of water molecules, water would behave much like many other liquids and, as a consequence, everything about life on Earth would be very different. However, if you bend your elbow so the angle is reduced to 104.5 degrees, you have a sense of what the water molecule actually looks like, and at this angle, water behaves in a profoundly different way from many liquids.

For example, water is denser than virtually any other naturally occurring substance that is a liquid at room temperature with the exception of mercury. Also, because it has a very high specific heat, it takes a great deal of energy to increase its temperature. This is why lakes and oceans have a moderating effect on their surrounding environments—the water is able to absorb immense amounts of heat with only slight changes in temperature, reducing the rise and fall of the temperature of the air near shore. Water has an enormous latent heat of vaporization, which is why, even over a blue flame in your kitchen, you can boil water in a pot for quite some time before seeing any noticeable change in the water level. It is also why perspiration cools your body so effectively. Furthermore, water has one of the highest surface tensions of any liquid with, again, the exception of mercury, explaining why so many insects are able to walk, stride, and skip about on the surface of water, something we will examine in the next chapter. All of these unusual characteristics of water are primarily due to this angle—the geometry of this extraordinary molecule.

To understand why the geometry of the water molecule has such an impact on how water behaves requires an understanding of an atomic property called electronegativity, which is essentially the ability of an atom to attract electrons to itself. The electronegativity of oxygen is high, while hydrogen's electronegativity is relatively low. Therefore, the electrons that oxygen and hydrogen share in the water molecule reside closer to the oxygen atom than to the hydrogen atoms. If the water molecule was not bent, this would not have a great impact. But, because it is bent, the molecule is not symmetric, and the more negative oxygen atom and the more positive hydrogen atoms together form a dipole. This changes everything—it means each water molecule has two sides—the side near the oxygen atom being negative and the side near the hydrogen atoms being positive. This dipole causes a net attraction between water molecules, the more negative oxygen-side of one water molecule attracted to the more positive hydrogen-side of another water molecule. This self-attraction, the desire of one water molecule to stay close to another water molecule, explains many of the characteristics of water described above. Its high surface tension is the consequence of all of these water molecules not wanting to leave each other—a drop of water wants to stay balled up and not spread out like it would if its surface tension were low. It takes more energy to boil water because boiling is the transformation of liquid water to vapor phase water, requiring the separation of one water molecule from the other, something they are loathe to do. Water is a simple little molecule with only three atoms, but because it is bent, its behavior is unique and complex.

The molecular structure of water is also responsible for one of the most spectacular aspects of water, the fact that ice floats. As liquid water transforms into ice, it forms a crystalline structure wherein the molecules orient themselves so the positively charged hydrogen side of one molecule is oriented toward the negatively charged oxygen side of another molecule. The resulting crystalline structure opens up some space between the molecules that wasn't there when it was in the liquid phase. This makes the density of ice less than that of its liquid counterpart. Therefore, it floats.

It probably doesn't seem odd that ice floats. The behavior is so embedded in our everyday experience that we don't appreciate how special it is. Get a glass of water with ice, and the ice is on the top of the glass, not the bottom. Icebergs float. Ice sheets float. When a lake freezes, it "freezes over," it does not "freeze under." This seems to us like normal everyday behavior. But this is not at all the case for most other substances. If you melt a metal and then let it refreeze, the solid appears at the bottom and edges of the container; the system will solidify from the bottom up. When you melt paraffin wax, the liquid will go to the top and the solid portion will stay at the bottom. Even carbon dioxide gets denser as it solidifies, dry ice sinking when you place it in a pool of liquid carbon dioxide.

The fact that ice floats on liquid water is not just some physical curiosity, it has enormous implications for life on Earth, including lake life. Without this characteristic of water, lakes would freeze solid, eliminating virtually all life in lakes. Or, perhaps, confronted with a completely solidified lake environment, a dramatically different evolutionary sequence would have occurred in lakes, resulting in fish and other species possessing the antifreeze-like properties that currently enable a handful of frog species to freeze and then reanimate in the spring (the best known of these is the wood frog of Alaska). But lakes do not freeze solid, or at least do not do so very frequently, and never freeze solid if they are sufficiently deep. Let us look at the process of lake freezing as winter approaches.

Away from the equator, as winter approaches, the air cools down relatively quickly, resulting in lake temperatures that are higher than the air above. And so the lake slowly cools by transferring heat from the lake surface to the air. Let us imagine what would transpire for the *imaginary* situation where water behaves like most other liquids. If this were the case, as the surface water cools, it would get heavier and would sink to the bottom of the lake. This process would continue as the water temperature continued to drop, the heavier surface water falling, and being replaced with relatively warmer water from below, which would in turn be cooled. In other words, the lake would mix, and it would continue to mix as it cooled. Because the lake would be well-mixed, its overall temperature would drop almost uniformly without large temperature differences between the top and bottom layers. So, as the overall temperature approached 32 degrees F., the lake as a whole would freeze, large crystals perhaps forming at the top and sinking, or maybe forming from the shore area and moving inward. It would depend on the details of the lake and the rate of cooling. But, the main point is that the lake would completely freeze. Every part of the lake from the very bottom to the very top would be ice.

Happily, this isn't what actually happens, and the reason for this has to do with how the density of liquid water changes as the temperature drops. Like most liquids, the density of water gets larger as you cool it, at least for most of its liquid phase temperature range. For most liquids this statement is true from the boiling point all the way down to the freezing point. But, for water, this is true only from the boiling point to 39 degrees F., after which the density actually decreases as you go from there to the freezing point. This is shown in Figure 6 where the density of water, ρ is plotted against temperature T for temperatures ranging from 30 degrees F. to 50 degrees F.

As Figure 6 shows, from 50 degrees F. down to 39 degrees F., the density rises to its maximum, after which the density falls as freezing is approached. The consequence of this is that, above 39 degrees F., as a lake cools at the air/water interface, the water cooled there will get heavy and sink. It will be replaced by relatively warmer water, which in turn will cool and sink. In other words, the lake will be fairly well-mixed, just like for the artificial case envisioned above, where the density of water increases all the way down to its freezing point.

But, in the actual case for water, once cooled down to 39 degrees F., further cooling occurs in the opposite way. As heat is removed at the air/water interface and the surface temperature is reduced, the water gets lighter, and so it just stays in place. In other words, the lake becomes stably stratified, the phenomenon we visited in our earlier discussion of overturning. Also, as the air continues to cool the water, this cooling effectively occurs only near the surface since there is no mixing of the lake. This will continue until the surface freezes over, leaving a lake that is at 32 degrees F. at the surface and close to 39 degrees F. in its hypolimnion.

Further cooling can and does occur if the air is lower than freezing. Heat is conducted away from the lake through the existing ice sheet,

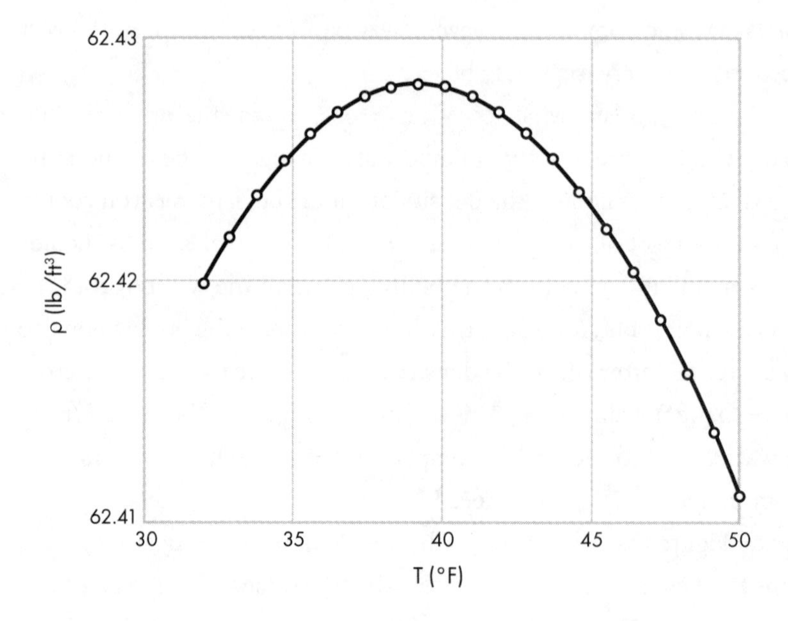

Figure 6. Plot of the density of water versus its temperature, showing a maximum at 39 degrees F. (data from the *Handbook of Chemistry and Physics*, 52nd Edition, 1971–1972, Chemical Rubber Company, Cleveland, Ohio).

and since the water at the ice/water interface is effectively at 32 degrees F., further cooling results in a progressive thickening of the ice sheet. This effectively protects the warmer body of water in the hypolimnion, which may be at or close to 39 degrees F., from getting any colder. Of course the presence of freezing water at the ice/water interface can and does cool the hypolimnion. But this rate of cooling is not rapid—the lake is stably stratified. Once covered by ice, wind cannot cause mixing, and so the transfer of heat from the hypolimnion to the cooler water above does not occur very quickly, effectively protecting the hypolimnion.

A temperature of 39 degrees F. is not something we would find comfortable. But for organisms living in lakes in northern climates, it works. Prior to freezing over, the lake probably was saturated with air. Due to the increasing solubility of oxygen with decreasing temperature, the amount of oxygen in the hypolimnion can be high. Hence, everything from fish down to single celled organisms can survive the winter, basking in the relatively warm water and breathing easy.

Of course things don't always go this way. Even covered in ice, heat continues to leave the lake. If, for example, the bottom of the lake is at 39 degrees F., and the air above is at a frigid 15 degrees F., the temperature difference will cause heat to flow from the lake toward the air thereby cooling the hypolimnion and thickening the ice sheet. For deep lakes, this is not a problem, but for shallow ones, the entire lake may freeze. And so, in colder climates, shallow lakes may be devoid of fish, the fish having died during a particularly cold winter in the past when the whole lake froze. Also, for relatively small lakes, even if they do not freeze solid, they can run out of oxygen if the lake remains frozen for sufficiently long. The amount of oxygen present is fixed, and fish aren't the only consumers. In lakes with a significant sediment bed, aerobic bacteria will be breaking down the organic detritus at the lake bottom, and if the mass is large, oxygen may be depleted from this activity alone, resulting in fish kills. But for large, deep lakes, survival is more likely.

It is often stated that ice serves as an insulator for a lake, but this is true only in a loose sense. In point of fact, ice conducts heat more readily than liquid water does, about three times better, and so, from this perspective, heat should leave a lake more readily once a lake freezes. But the real point is that, once a lake is frozen over, the liquid water is protected from wind. It isn't so much that the lake is *insulated* from the air above, but rather that it is *isolated* from it. Once frozen over, the lake is protected and the hypolimnion, in particular, is able to stay at a relatively warm 39 degrees F. It should be noted that a lake *can* be well-insulated from the air once snow falls on the ice surface. Although the thermal conductivity of snow varies with its structure, it insulates better than both liquid water and ice and so once a lake is frozen *and* covered with snow, it is quite well-insulated from the cold weather above.

• • •

Once frozen over, a lake is a very different place, not just below the ice, but also on top of the lake and along its shore. Ice on a lake is like a living, breathing thing. It constantly grows and shrinks, a process governed by the daily heating/cooling and melting/freezing that result from the changes in air temperature that accompany the rise and fall of the sun as well as by storms coming and going. In large lakes, tides can play a role. All of this expansion and contraction can cause a range of sounds lakeshore residents in cold weather climates are familiar with: one may hear everything from grinding sounds to sharp cracks, as well as thunderlike booms, squeaks, and groans.

The growth of the ice sheet can do more than make noise. Drops in air temperature can cause the ice sheet to contract, forming cracks at various points that fill with liquid from the water beneath. This water will subsequently freeze, filling in the cracks, and resulting in a net horizontal growth of the ice sheet, a growth that is enhanced when the temperature rises and the sheet expands. An expanding ice sheet has nowhere to go but outward, up onto the lake shore. This expansion is referred to as *ice heaving* or *ice jacking*. On large lakes the force of this expanding sheet of ice can be significant, causing what are called *ice ridges* or *ice pushes* along the shoreline, gouged-up lines of earth as high as 5 feet tall, all the result of the push of the expanding ice. These phenomena are the bane of lakeside property owners because ice activity grinds up the shoreline area, can damage decks and other structures near shore, and may even push houses off of their foundations. Similar effects can occur when wind blows ice onto shore after the ice is partially broken up. Once open water exists somewhere, the wind can push a sheet of ice into the shore, which can also cause enormous property damage.

• • •

Once frozen over, the human experience of a lake is a decidedly one-sided affair. During the summer we may swim in a lake, or dive beneath it—we can experience its surface, its shore, and its depths. Although the ice on a lake provides one with the opportunity to walk upon it, to cross it without need of a boat or the athletic ability of a swimmer, ice also makes the water beneath distant and inaccessible. The ice sheet separates humans from the lake in a way that makes the water almost unknowable. It is as if the liquid water of a lake in winter is transported to a different part of interstellar space, known only to the ice fisherman who, huddled over a hole, experiences the liquid water through his small circular porthole, this wormhole that occasionally transports a fish up from the depths. It is known only by highly trained scuba divers, rescue divers and scientists who, like astronauts, venture into the dark winter realm, using lights to turn the blackness into a murky dark blue.

• • •

There is a distant possibility that there is another way to experience life beneath the surface of a frozen lake, a possibility I have spent an inordinate amount of time thinking about—one that stems from a work of short fiction. "The Hermit's Story," by writer and environmentalist Rick Bass, describes the experience of Ann, a hunting dog trainer and her client, Gray Owl. It is the middle of winter, and Ann has delivered several dogs she has trained for him to his home in rural Saskatchewan. Out in the field, while acquainting Gray Owl with his newly trained canines, Gray Owl falls through an ice-covered lake. Fearing the worst, Ann carefully ventures to the edge of the hole, only to find Gray Owl dry and unharmed at the bottom of the lake. "This happens a lot more than people realize," Gray Owl tells Ann. "A cold snap comes in October, freezes a skin of ice over the lake—it's got to be a shallow one, almost a marsh. Then a snowfall comes, insulating the ice. The lake drains in fall and winter—percolates down through the soil . . . but the ice up top remains." The remainder of the short story is nothing short of magical. Somewhat lost, Ann and Gray Owl decide to walk the lake bottom, believing its far shore will be near a road. Night falls and they are able to see the stars and then the moon filtered through the sheet of ice above in colors of cobalt and silver and blue. They walk beneath the ice, lighting their way with burning torches made of cattails that occasionally ignite pockets of methane, creating puffs of flame. It is one of those beautiful, imaginative stories that makes you wish you could live in it, that you too could find this drained, frozen, lake where you could walk about bathed in ice-filtered moonlight while listening to the squeak and groan of the ice sheet above you.

Of course "The Hermit's Story" is a work of fiction and one mustn't dwell on such happy fantasies too long. But, when the story was published in *The Best American Short Stories 1999*, Bass revealed in the Contributors' Notes that he'd gotten the story idea from a Frenchman who claimed the Canadian writer Grey Owl (presumably also the source of the character's name, though with different spelling) once found a frozen lake with no water in it. As far as facts are concerned, this is all that exists, to the best of my knowledge. But I've been searching for evidence that such a lake can exist ever since. E-mails to noted limnologists have produced little more than skepticism. And Grey Owl, though a noted conservationist who died in 1938, turns out to have been something of a fraud, having posed as a First Nations person in Canada when he was actually an Englishman, Archibald Stansfeld Belaney. Still I search, hoping that my love of lakes can extend to their depths in the depth of winter, and that one day I too may walk a lake bed, viewing the moon through a sheet of ice, wearing dry boots.

11. SURFACE TENSION

In the previous chapter we noted how the unique geometry of the water molecule accounts for many of its interesting qualities. Notable in that discussion was the dipole that results from the large electronegativity of oxygen, creating a net attraction between water molecules. It is this attraction, this desire of water molecules to stay close to each other, that creates surface tension. It is this surface tension that causes water drops to tend toward a spherical shape, making water drops "bead up" on the surface of a car or a window. It is surface tension that causes a meniscus to form when a liquid is held in a container. Surface tension is the source of the capillary forces that are partially responsible for the ability of trees, even giant sequoias, to pull water up from the ground all the way to the tree crown. Because the intermolecular forces between water molecules are so large, the surface tension for water is among the largest of known liquids.

Surface tension is measured in units of dynes, a unit of force, per centimeter. This unit, dynes/cm is the standard unit for surface tension and provides a measure of how much force a meniscus can exert. For example, inside of an air bubble situated in a liquid, there is an excess pressure due to surface tension. This pressure is equal to twice the surface tension divided by the bubble radius. The smaller the bubble radius, the bigger is this pressure. Also, the larger the surface tension, the greater the pressure.

Perfectly clean water has a surface tension of 72.1 dynes/cm. To put that number in context, acetone has a surface tension of only 26.2 dynes/cm, ethanol is 24.0 dynes/cm, isopropyl alcohol is 21.7 dynes/cm, and octane, the main component of gasoline, is 21.8 dynes/cm. There are substances whose surface tension exceeds that of water, but they are rare. One is mercury, whose surface tension is an incredible 486 dynes/cm; another is hydrazine, a rocket fuel, which has a surface tension of 91.5 dynes/cm. But mercury and hydrazine are oddities. When scanning through a table of liquid properties, water tends to be close to the very top when it comes to surface tension. It seems water, this most common of liquids, has a surface tension exceeded by only the most exotic of substances.

The effects surface tension has on a liquid are many. One is the fact that it causes a sample of liquid to minimize its exposed surface area. This is why a drop tends toward a spherical shape, a sphere being the geometry that minimizes the surface area of a fixed mass of liquid—hence the spherical shape of raindrops. But surface tension doesn't act in a vacuum, and the way in which it manifests itself is due in part to how it interacts with the other forces at play. One of these is weight. Although water tends toward a spherical shape, the water contained in a drinking glass isn't spherical in shape because the down force of its own weight nestles it firmly into the glass, forcing the water to take on the shape of the glass. It is only near the surface that surface tension plays a role, causing the air/water interface to have a curved meniscus near the glass. The relatively small role surface tension plays in a glass of water is the result of the large mass of water present.

If, on the other hand, the mass of water present is much smaller, then gravity will play a commensurately smaller role. Dip your finger into that glass of water and dab the result onto your kitchen counter, and the water will conform to a roughly hemispherical shape. Here, gravity is pulling the surface down, while surface tension is trying to bead the water up into a sphere. A compromise between the two forces is reached, the result being a somewhat hemispherical shape.

We can make the role of gravity even smaller by eliminating it. This is what happens during experiments with water drops on space platforms such as the U.S. Space Shuttle and the International Space Station, where water drops larger than a grapefruit in size can be created. Images and videos of these experiments are easily found on the internet—it's fascinating to view these large spherical drops, spherical because the only force present to determine their shape is surface tension.

There is one more factor that comes into play, something called "wettability." Take two identical drops of water—place one on the surface of a Teflon-coated pan and the other on the surface of a non-Teflon surface (say, the metallic back side of the same pan), and the difference is obvious. On the coated surface, the drop beads up tightly, appearing almost perfectly spherical, while on the metallic surface the interface is hemispherical. Both drops have the same surface tension and weight. So what causes the different shapes? The answer is the wettability of the solid surface. Though water is highly attracted to itself, it always has some finite degree of attraction to the solid surface upon which it resides. For surfaces referred to as hydrophobic, that attraction is very small. The Teflon surface described is considered hydrophobic, so water beads up tightly on it. For surfaces which are referred to as hydrophilic, water is somewhat attracted to them and spreads out upon them. The water drop placed on the metallic back surface of the pan attains a hemispherical shape, the water somewhat attracted to the metal and somewhat attracted to itself. So, it is this interplay between weight, surface tension, and wettability that determines how water behaves when deposited on a solid.

But what does any of this have to do with lakes? The answer is obtained if we imagine flipping the system upside down and placing something solid on the surface of water, on a lake surface, for example. When we do this, it is still weight, surface tension, and wettability that determines what happens, except now we are determining whether the solid object will float or sink—for example, whether a leaf floats or sinks when it falls on the surface of a lake. It is these factors that explain why pollen floats and sand sinks. When a large mass is placed on the surface of water, it will simply sink, regardless of surface tension or wettability; this is why a large rock placed on a water surface promptly sinks to the bottom. Objects that are less dense than water, such as a wooden plank, will float. In both cases, surface tension plays a role, but it is a minority role—it provides an upward-directed force, but it is small compared to the downward force acting on the rock, or the upward-directed buoyancy force acting on the wood.

However, as the overall length scale of an object gets smaller, the gravitational force decreases as well, and surface tension and wettability become more important. This length scale at which surface tension begins to dominate is typically on the order of a millimeter for water (about 0.04 inches), though this depends on the density of the solid object. Nevertheless, at roughly that scale or smaller, objects denser than water may in fact float due to water's surface tension force. Indeed, very small bits of rock dust can and do float on water, depending on their wettability. Insects and spiders tend to be millimeters in size, and indeed many of these creatures have densities greater than water. It is the combination of the surface tension of water and the wettability of these creatures (or lack thereof) that allows some of them to float on water. These forces also enable water locomotion, the fascinating process by which certain creatures don't just float but can in fact walk on water.

There are a surprising number of creatures (over 1200) that can literally walk on the surface of water. While insects such as water striders are the most well-known, there are also spiders, lizards, birds, and even mammals that can effectively do this. Some examples include the basilisk lizard, the western grebe (a bird), and dolphins (which, incredibly, are able to "tail walk" across the water surface). However, the vast majority of water walkers are arthropods (spiders and insects). For these relatively small creatures, it is again the interplay between their weight, surface tension, and wettability that determines whether and how they walk on water.

When a water strider floats on water, its weight is supported by its legs, long thin rods that tend to lay flat on the water surface, like tiny floating logs. The weight of the insect causes these legs to deform the water surface, indenting the surface and increasing its total area. This increase in area is in direct opposition to what the water molecule wants; as noted above, surface tension acts to minimize the surface area of any body of water. Hence, when the water strider's leg deforms and increases the area of the water surface, that leg is met with an upward-directed opposing force—the surface tension force. It is as if a thin elastic membrane exists on the water surface, and the more it is deformed, the more it resists deformation, supporting the weight of what resides upon it. As long as this upward-directed force exceeds the weight of the water strider, it will stay above the water surface. Furthermore, to move on the surface of water, those legs have to accelerate backwards in order to propel the insect forward. This results in an additional force, but if the combined locomotive forces and weight force are less than the surface tension force, the water strider's leg will remain at and above the water surface, enabling it to float and move about.

The process can be fraught. Should the water strider's leg penetrate through the water surface, a significant "energy penalty" is incurred in the process of pulling the leg back out. So it behooves these creatures to avoid this. Since the surface tension of water cannot be changed, these creatures have evolved coverings on their legs to increase their hydrophobicity, that is to say, they decrease their wettability. A combination of miniscule hairs and grooves on their legs, as well as chemical secretions, make their legs virtually nonwettable, a property called superhydrophobicity.

Superhydrophobicity is a pretty incredible phenomenon in and of itself and is not uncommon in nature. In the plant world, leaves such as those of the elephant ear, the lotus, and the rice leaf, all exhibit superhydrophobicity. Place a drop of water on any of these leaf surfaces and it will roll around like quicksilver or a puck on an air hockey table. Unlike their more wettable counterparts, these leaves are never really wet. The water rolls right off, no matter how much you apply. Shake one of these leaves after a rainstorm and droplets roll off completely, leaving behind a perfectly dry leaf. In the plant world, this enables excellent self-cleaning properties, the water drops taking bits of dust and detritus with them as they go. Engineers and scientists have sought to mimic this capability to create, for example, self-cleaning windshields, paints, solar panels, and other engineered surfaces.

Superhydrophobicity is defined as the condition where the contact angle between a solid surface and the water surface is greater than 150 degrees. If a water drop were to not wet a surface at all, it would have an infinitesimally small spot in actual contact with the surface. Under these conditions, the contact angle would be a full 180 degrees. Hence, a contact angle of 150 degrees is not far from what one obtains for something that does not wet the surface at all.

Studies of the water strider show the contact angle for their legs surpasses the minimum superhydrophobicity requirement at a whopping 167.6 degrees. The superhydrophobicity of the legs of these insects was for some time thought to be due entirely to a secreted cuticle wax. However it has been shown that the cuticle wax by itself yields a contact angle of only 105 degrees. The superhydrophobicity is actually provided by microscale and nanoscale hairs and other structures on the leg. Air gets trapped in between these structures, significantly reducing the actual water/leg contact area. Effectively, a water strider leg is surrounded by an air cushion (a "plastron"), explaining its ability to move about without sinking.

The ability to avoid wetting enables all manner of insects to manipulate the water surface in various ways. Springtails, a type of surface-dwelling insect, are able to use their bodies to curve the air/water interface, effectively storing energy like a spring. They are able to suddenly release this stored energy, leaping vertically upward, a mechanism they use to evade predators. Other insects are able to submerge themselves, encompassing their entire bodies in their air plastron, enabling them to reside beneath the water while breathing and to then return to the surface, all while staying dry. One must be envious of these insects, these water striders and the like. Able to scoot along at speeds in excess of 4 feet per second, they live on a virtual trampoline, a soft world whose surface they can manipulate to their own desires, using this thing called surface tension.

But perhaps the most interesting thing about the surface tension of water isn't its magnitude, but rather the fact that it is not a constant—that it varies. Surface tension is indeed a tension, which is to say a force. Thus when tension varies from one location to another, there is a force differential between those two locations. When two forces are equal, the situation is static. This is the situation you have when an object is placed on a table. The down force of the object is exactly balanced by a supporting force from the table. But when there is a differential in forces, acceleration results. And, when the force differential occurs in a fluid, the result is a flow, or, on the surface of a liquid, the result is a surface flow.

An example of a surface flow created by a difference in the surface tension of water can be seen when you drop an ember on a water surface. Back in the days when smoking indoors was accepted and popular, this phenomenon could be observed whenever you flicked a still-lit cigarette into the center of a toilet bowl. The result was a rapid surface flow outward from the location where the cigarette landed. The burning ember, well-endowed with dark ash, visualizes this flow via radial streaks of black ash, well-highlighted by the white porcelain. Why does this occur? Surface tension varies significantly with temperature, decreasing as the temperature rises. So, on the water surface, the region where the hot cigarette ember lands is heated, and the surface tension at that location quickly drops. Away from the ember, the water surface remains cool, nominally at room temperature, thus the surface tension there is high. A flow is formed moving from the region of low surface tension to high surface tension, a flow that continues as long as the temperature difference is maintained. For the cigarette ember example, this flow is brief, lasting only until the ember is quenched and cooled by the water, perhaps a second or two. These flows are referred to as surface-tension-driven flows, or Marangoni flows (named after Carlo Marangoni, the Italian physicist who studied surface phenomena in the mid-1800s).

And, it isn't just temperature that is capable of altering the surface tension of water. One of the most interesting things about the surface of water is that it is almost never occupied by pure water molecules. Not even close. Water is almost always covered with a monolayer of some organic compound or another. These compounds are often "surfactant monolayers" (the word surfactant is a contraction of the phrase "surface active agent"). Surfactant monolayers are indeed monolayers, layers that are a single molecule thick, and though something that thin does not have a great deal of mass, it nonetheless has a major impact on the water surface by reducing its surface tension. You are familiar with surfactants in the form of soaps and detergents that reduce the surface tension of water thereby facilitating its ability to clean fabrics. But, many things can form a surfactant monolayer on water. Indeed, as we will see below, obtaining a water surface that does not have a monolayer on it is virtually impossible.

But the really interesting thing about surfactants isn't that they reduce the surface tension of water, but rather that they do so in proportion to their surface concentration. That is, the more surfactant present on the water surface, the lower the surface tension gets. This probably makes intuitive sense, though a consequence of this straightforward fact is a bit trickier to understand. But understand the consequence we must, because it is important to bodies of water ranging from livestock ponds to oceans. If, as noted above, the surface tension decreases as the surfactant concentration increases, then it must be the case that if one artificially increases the concentration of the surfactant layer by compressing it in some location, then, the surface tension would be lower in that location and higher everywhere else.

Imagine we have a small tank filled to the brim with water. If the water is obtained from the tap, then there will certainly be a monolayer on it (more on this, below). Now, let's imagine that we lay a rod or dowel down upon the tank, one long enough to span the width of the tank with length to spare. Let's imagine the rod is situated in such a way that it divides the water surface roughly into two halves. If we now take this rod and slide it across the tank surface (without spilling any water), decreasing the area of the portion of the surface ahead of the rod and increasing the area behind, we will be increasing the surfactant concentration on the water surface ahead of the moving rod, and decreasing it in the region behind the rod. Therefore, the surface tension on the side of the rod with the low concentration will be high, and the surface tension on the side of the rod with the high surfactant concentration will be low. In other words, there will be a force differential that will seek to move the rod in the direction opposite from the direction we are pushing it, a force seeking to return it to its original position. The more we push, the greater this force will be. This increase in force with increasing concentration of the monolayer is modelled by surface scientists as an elasticity since the surface behaves as if there were an elastic film covering the surface (one that resists the motion of the rod). Although the presence of actual rods on the surface of lakes is uncommon, this elasticity nevertheless has a profound effect on processes occurring above and below the water surface.

The air/water interface of a lake or ocean is the place that mediates the transport of many important things. Dissolved gases such as carbon dioxide, oxygen, and methane move to/from air to water at this interface. This is not a small thing given that the ocean is the largest single sink for human-created carbon dioxide. The more effectively transport of gasses across the air/water interface occurs, the greater the carbon dioxide uptake by the world's oceans and lakes would be. In the absence of a surfactant monolayer, water flow beneath the air/water interface would be relatively unimpeded. For example, an upward-moving plume rich in oxygen and devoid of carbon dioxide would rise to that surface, exposing the oxygen-rich water right at the interface, enabling the air to actually touch it, absorbing the oxygen and transferring carbon dioxide into the water. This would be a very efficient means for the exchange of these gasses.

But in the presence of a surfactant monolayer, the abovementioned elasticity significantly impedes all of this. Subsurface water rising up and deflected outward is resisted in its flow by the elasticity of the surface. This has a big impact. It has an impact on gas exchange, on heat transfer to and from the air and water, and it has a major impact on evaporation. All of these processes are impeded by the elasticity of monolayers, resulting in reduction in gas exchange and heat transfer and evaporation.

It should be noted that all of this activity is perfectly natural. Surfactants come from the body fats of all kinds of living organisms, from the oils generated by plants—virtually any organic molecule will spread upon a water surface and generate some degree of elasticity. So, it is not as if some change in human behavior would eliminate surfactants from lakes and oceans, enabling greater transfer of carbon dioxide from the air to water. Nevertheless, it is stunning to think that something so thin, literally a mere molecule in thickness, a layer thinner than a typical cell in your body, has such a dramatic impact.

Nor are surfactant monolayers some kind of rarity. In fact, the exact opposite is true; it is almost certainly the case that you have never seen nor touched a water surface that did not have a monolayer on its surface. Lakes, oceans, rivers—they all have surfactants on their surface. Water from the cleanest of taps, when poured into a glass, will promptly form a monolayer within seconds. In the research I have done on the role of surfactant monolayers on air/water transfer processes, in order to quantify the effect of a surfactant, one must run a control experiment, that is an experiment without a monolayer—a "clean" case, completely devoid of surfactant monolayers. Such experiments can be fantastically difficult. Even when using doubly distilled water, ultraviolet filtering, and tedious surface cleaning procedures, the ability to create and maintain a clean water surface is extraordinarily challenging. This is because almost any kind of organic molecule will form a monolayer on water. If you were to create a perfectly clean water surface and simply touched the surface with the tip of your pinky, a monolayer would immediately form. Describing his own laboratory efforts in his 1974 Scientific American article, "The Top Millimeter of the Ocean," Ferren MacIntyre of the Graduate School of Oceanography at the University of Rhode Island, confided: "Just once have I been able to keep a surface clean for 24 hours. It took a year of preparation, and even then it was mostly good luck, since I never succeeded again."

We have had more success in my laboratory due in part to the availability of more modern methods, including the use of infrared imaging for the detection of surfactant contamination. But, the main point is that if you have in front of you a lake, a pond, a reservoir, or an ocean, you also have before you a surfactant-covered water surface.

Surfactants are invisible and affect surface tension in myriad ways. They are everywhere. And they are powerful.

12. EVAPORATION

At some point in your high school education, perhaps in a chemistry or physics class, you were probably told, "evaporation is a cooling process." It is. And that is good. Evaporation of your perspiration cools your body, enabling you to maintain your temperature in just the right range. But evaporation is also a mass loss process, and just as evaporation of your perspiration without replenishment will eventually result in your demise, so too can the water lost by evaporation from a lake result in the effective elimination of a lake or reservoir, a thing of some concern in locales where a lake is used to provide water for human use. Evaporation is invisible to us, the molecules of water on a lake surface transformed from the liquid to the vapor phase, silently drifting into the atmosphere where they may fall again as rain, typically very far away, and often where the droplets do little to help replenish lakes and reservoirs, as when rain falls over the ocean.

Evaporation is a big deal. Even in wet climates, such as South Carolina where I live, a lake can lose over 50 inches of water to evaporation in a year. Near my home there are three interconnected reservoirs, Lakes Hartwell, Russell, and Thurmond, each formed by a dam on the Savannah River, the river that forms the border between South Carolina and Georgia for most of its length. Lake Hartwell experiences an average annual evaporation rate of about 52 inches per year, which is likely close to the average for the other two reservoirs. If you take this evaporation rate and multiply it by the surface area of just these three reservoirs, you get 661,400 acre-feet of evaporative loss per year. That volume is equivalent to slightly less than 29 billion cubic feet, or 216 billion gallons. That is a lot of water. It is equal to 11 percent of the total volume of these reservoirs when they are full. At Clyo, Georgia, near where the Savannah River enters the Atlantic Ocean, the total annual outflow is 8.2 million acre-feet/year, meaning the loss due to evaporation, just from these three reservoirs, is equivalent to 8 percent of the outflow of the entire Savannah River.

The above example pertains to the wet southeastern United States. If we now look at the much drier western part of the United States, where relative humidities are much lower and evaporation rates much higher, the impact of evaporation is even more dramatic. Indeed, in the 17 western states, annual evaporation from impoundments is 15.6 million acre-feet, an amount comparable to all of the water stored in all of the reservoirs of California (in 1965). Lake Mead by itself, which provides water for Las Vegas and several other cities, and where the annual rainfall is a paltry four inches, experiences an evaporation rate estimated at 76.0 inches per year. Multiplying this rate by the surface area of Lake Mead gives a volume of 1.0 million acre-feet/year lost to evaporation. By way of comparison, the entire state of Nevada is permitted to remove 300,000 acre-feet per year from the Colorado River. In other words, astoundingly, the amount of water lost to evaporation from Lake Mead is over three times larger than what Nevada is allowed to remove from the Colorado River.

Given the large amounts of water lost to evaporation, it is not surprising that valiant attempts have been made to reduce or eliminate evaporation from lakes and reservoirs, particularly in the West. In drought-riven years, water managers in dry climates have resorted to desperate measures. In one well-publicized situation in 2015, the Los Angeles Department of Water and Power, in response to a drought, dropped 96 million floating black balls called shade balls onto the surface of a reservoir in an effort to reduce evaporative loss. The balls covered the water surface leaving little space in between, serving to protect the water from solar heating and reducing the flow of water vapor from the water surface, thereby reducing evaporative loss. The program was implemented on several reservoirs in the Los Angeles water system, and images of the 4-inch diameter plastic balls covering these reservoir surfaces became a brief media sensation. Though the shade balls on some of these smaller Los Angeles reservoirs have since been replaced by floating covers, the large 175-acre Los Angeles Reservoir continues to be protected by shade balls, where they have reduced evaporation, by 85 percent to 90 percent, an effective solution if not a cheap one.

Another way to reduce evaporation is to simply prevent an air/water interface from forming in the first place. Although reservoirs are a staple of water-resources management, there are other ways to store water, namely by taking available water and pumping it into the ground. There, safely contained in an aquifer, evaporation cannot steal water away. Indeed, by pumping water down through porous rock beds, and then pumping it back up some distance away, water is not just protected from evaporation, it is made cleaner, the porous rock serving as a filter, removing pollutants and microorganisms.

But the existing water system in the United States already relies heavily on reservoirs. Lake Powell and Lake Mead and the thousands of other reservoirs used throughout the nation represent an enormous taxpayer-financed investment in large infrastructure, and the water providers who use them are not going to abandon them any time soon. Furthermore, the population of the United States seems to continue its ever-southward migration with well-watered cities such as Detroit, Chicago, and Buffalo losing people each year to dry southern and western locales such as Dallas, Phoenix, Las Vegas, and others. In the Great Lakes, the United States is endowed with one of the largest lake systems in the world, a truly massive volume of fresh water. Yet, as a population, we are fleeing that wondrous supply of water at breakneck speed. So, like it or not, the reservoirs we've built in the South and West are not only important, they are becoming more important with each passing year.

Periodically the question arises: what can be done to eliminate or reduce the large evaporative losses that reservoirs experience? This question is ignored during years of water plenitude, but returns during drought years. It certainly gains more urgency as droughts become longer, more frequent, and more severe as time passes.

Imagine you are the head of a local water provider and you are standing on the shore of your city's reservoir. It is July, it is hot, and the reservoir level is low. The water surface lays beneath the hot sky, evaporation occurring unimpeded, the reservoir measurably dropping each day. "Isn't there a way to turn off this evaporation?" is surely the question that runs through the minds of those responsible for providing water to cities and municipalities during droughts. Short of spending millions of dollars to manufacture black shade balls, is there some other way to prevent all of this water from leaking up into the sky? Is there some kind of skin or film that could be spread across the immense area of a reservoir to hamper or stop evaporation and protect the water, at least until the rains come? Though riddled with caveats, the answer to that question is actually yes.

In the last chapter we examined surfactant monolayers, these single-molecule-thick films that water is never free of. Though just one molecule thick, they have a powerful impact on surface tension and can imbue the water surface with an elasticity that is otherwise absent and that profoundly affects how water absorbs heat and dissolved gases. But under the right conditions, certain classes of surfactants can do even more—they can also effectively stop evaporation. This is especially true of monolayers referred to as solid-phase surfactant monolayers.

Surfactants have a molecular structure that consists of long hydrocarbon chains. For many surfactants these long chains show no real structure on the water surface—they are more or less randomly oriented with respect to each other. However, solid-phase surfactant molecules orient themselves vertically, their chains perpendicular to the water surface. This is because one end of the hydrocarbon chain is hydrophilic and therefore wants to attach itself to the water, while the other end is hydrophobic and seeks to be as far from the water surface as possible. By pointing straight upward, both ends of the molecule get their way. In this configuration, the molecules are aligned parallel to each other like blades of grass on a lawn, each molecule packed closely to its neighbor. Viewed from above, the molecules are located in a grid pattern, forming a two-dimensional crystal. In addition to causing the film to attain a lateral rigidity, much like its three-dimensional cousin, this molecular structure provides virtually no space between the surfactant molecules. Water at the surface is forced to travel along the narrow passages between these long molecules, severely limiting evaporation. Hence, a solid-phase surfactant film, once formed, effectively shuts evaporation down.

The ability of solid-phase monolayers to reduce the rate of evaporation has been known for some time. However, a period of relatively intense research occurred in the 1950s and 1960s. The suppression of evaporation by monolayers was vigorously investigated by groups in the United States and Australia; laboratory studies and field campaigns were conducted to determine if and how suppression of evaporation by surfactant monolayers could be successfully implemented on lakes and reservoirs. Interestingly, or you might even say disturbingly, many such studies were conducted on reservoirs that were actively being used as the water supply for cities, making guinea pigs of the residents.

Let's look at what was done, and learned, here in the United States. Many of the evaporation-suppression studies were conducted by the Bureau of Reclamation. Especially notable was the deposition of monolayers on Lake Hefner, a 2500-acre lake in Oklahoma City, which is part of that city's water supply. Experiments were conducted on the lake for 86 days using a surfactant mixture composed primarily of hexadecanol with smaller amounts of tetradecanol and octadecanol. These surfactants are relatively harmless and are commonly used in consumer products such as shampoos, lotions, and cosmetics. Still, spreading these chemicals over a reservoir every day for 86 days, at times covering as much as 89 percent of the surface seems a bit risky. Nevertheless, this was done, along with similar studies at Ralston Creek Reservoir, a 150-acre lake that is part of the Denver water system, on a portion of Lake Mead, and on other reservoirs. The material deposited was applied in a variety of ways, often blown on as flaked material from a boat or sprayed as a slurry of the surfactant. A large amount was applied, considering the fact that the desired result was a layer one molecule thick. In the Lake Hefner study, 0.3 pounds of material were deposited per acre per day. This translates to an awful lot of material. If we use the 89-percent-coverage value cited above, this means deposition as high as 668 pounds per day. The average coverage on Lake Hefner was 10 percent, but still this corresponds to deposition of 75 pounds of material per day, adding up to 6450 pounds, slightly more than three tons for the 86-day study. It is easy to see that this practice would, if widely implemented on lakes throughout the west, lead to a huge consumption of materials.

In the laboratory, spreading a solid phase monolayer is relatively straightforward, and these monolayers do reduce evaporation dramatically. But the water surface one obtains in a laboratory tank is very different from that on a lake or reservoir. On such a large area, the monolayer is spread outward from many flakes or pieces of bulk surfactant (depending on how it is deposited). Where these individual monolayers meet, gaps inevitably form where evaporation is still able to occur. Moreover, in the outdoors, many factors cause the dissipation of surfactant films once formed. Ultraviolet degradation, bacterial degradation, and especially wind and waves, all serve to reduce the coverage. To effectively use surfactants to control evaporation would require continuous application potentially via boat or other movable platforms, or perhaps via multiple installed and partially automated devices. All of this equipment would require maintenance, upkeep, and resupply, again requiring boats. The whole scheme turns out to be daunting logistically and, inevitably, costly. Thus, not surprisingly, the use of surfactant monolayers to control evaporation has not been embraced and has not enjoyed significant implementation beyond field studies.

Today the use of surfactants in evaporation reduction is limited primarily to what are called *liquid pool covers*, surfactant solutions spread over a pool to reduce evaporation, thereby keeping the water warm. Though not employed in any significant way, the fact that monolayers were considered a potential method for reducing evaporation on reservoirs despite their many drawbacks reveals how truly constrained our freshwater resources have become.

● ● ●

But, do we really want to suppress evaporation in the first place? We began this chapter by noting that evaporation is a cooling process. So, without it, the temperature of lakes and reservoirs would rise. If we were able to suppress evaporation from reservoirs on a large scale, what would be the impact of the increased reservoir temperatures, particularly for very large reservoirs that host complex ecosystems? Perhaps the failure of surfactant monolayers to work effectively is the best thing that could have happened to the ecosystems of the lakes and reservoirs humans use.

And perhaps it is wise to step back even further and ask whether the formation of reservoirs was a wise thing to do in the first place. This is certainly an academic exercise given the prevalence and continuing construction of reservoirs of all sizes throughout the world. But still, it is a question worth asking. Evaporation shows that there is a bit of irony associated with reservoirs—we build them to hold onto the water that would otherwise flow via rivers out to the ocean. But, in creating these reservoirs we have also created enormous water surfaces, vast air/water interfaces that didn't exist before and where prodigious quantities of water are lost via evaporation, evaporation that didn't occur before they were built. Of course there is a net gain. The amount of water humans are able to withdraw and use from these reservoirs is much larger than would otherwise be available, evaporation notwithstanding. These reservoirs wouldn't have been built in the first place is this wasn't the case. But still, it forces us to ponder the results of all that extra evaporation into the atmosphere. What are the long-term consequences of this additional evaporation, and how can we even quantify it? If one searches the journal literature about reservoirs and evaporation, a whole host of articles focusing on the impact of climate change on evaporation from reservoirs are found. But there is a lack of research on the converse, that is, on how evaporation from reservoirs impacts and has impacted the climate. In holding onto all of this freshwater, we are also forcing a very large amount of it up into the atmosphere. One wonders what the consequences are and will be.

DEATH

13. DEATH BY HUMAN

Lakes are always in the process of dying and, ironically, for the very same reason that they exist in the first place: they are collection basins. Lakes are places where water collects, but they are also places that collect anything water can possibly carry. The streams, rivers, and shore runoff entering a lake carry with them all manner of things: sand, silt, plant matter, and much more. When these inflows slow down, as they necessarily do when they enter a lake, all of this matter settles out, the sediment load accruing year after year, decade after decade, making a lake shallower and shallower as time passes.

And it isn't just runoff. Productive lakes rich in plant and animal life are continuously depositing material onto the bottom. Photosynthetic life fixes carbon from carbon dioxide transforming the carbon into biomolecules. These biomolecules become part of everything from bacteria to bass to beavers, and these animals leave behind their bodies, their excrement, their skeletons, parts of which ultimately find their way to the lake floor.

Although lakes can die from more dramatic forces, like the collapse of an underground cavern that drains a lake in a matter of days, this tends to be uncommon, and many lakes simply die from the slow accumulation of matter. In a process often referred to as lake senescence, a deep clear lake becomes a shallow weed-filled lake, which perhaps becomes a wetland, and then a stream, until finally nothing that looks remotely like a lake is left. The range of time over which this process occurs varies enormously, literally from the course of a human lifetime for something like an oxbow lake (lakes formed in the abandoned bend of a river), to millions of years for tectonically created lakes. But still, all lakes are going through this same process of dying. The fate of a lake is perhaps best and most succinctly described by G. A. Cole in his 1994 *Textbook of Limnology*, where he remarks poignantly: "Once formed, they are doomed."

The amount of material transported into a lake can be massive. As just one example, the Great Slave Lake in Canada's Northwest Territories receives 100,000 tons of sediment from the Slave River each *day*, an extraordinary amount. The enormity of this number is somewhat mitigated by the fact that the Great Slave Lake is the deepest lake in North America, having a floor that reaches to 1522 feet below sea level, and a total maximum depth (the maximum distance from the surface to the floor) of 2014 feet. Still, this is a very large amount of sediment to be transported into a lake.

The material entering a lake can come from a variety of sources. Erosion of the lake shore by runoff and wave action undercuts the lake perimeter, transporting large quantities of rock and soil to the lake bottom. Winds provide material for a lake's end, the dust, soil, and organic detritus carried by even the faintest of breezes deposit on a lake surface and ultimately filter down to the bottom. And the rich habitat that lakes provide for all classes of organisms, from single celled life forms to fishes and mammals—all of these also deposit material in lakes, shellfish doing this in an especially effective way. It seems to be nature's way to reward a lake for all that it does by slowly and unceasingly acting to make it disappear.

In fact, the situation of lake senescence is a bit more complex. Indeed, the entire process by which a lake evolves is somewhat controversial, and for good reason. For example, some of the oldest lakes in the world are also some of the deepest, a fact that seems to be at odds with the idea of a lake getting shallower the more time has passed.

This calls into question the aforementioned simplistic notion of a lake's demise by progressive deposition of detritus on the lake floor and suggests that other factors may also contribute to the fate of lakes. Moreover, though the sediment load entering a lake can be large, when spread over the entire lake floor, the resulting thickness is not all that impressive. Indeed, typical lakes accrue sediment layers at a rate on the order of a millimeter per year. At this rate, a 100-foot-deep lake would require slightly more than 30,000 years to fill in, and a deep lake like Lake Tahoe would take over 300,000 years.

This slow pace complicates the story of lake senescence, given that the duration of interglacial periods within our current ice age is only on the order of 20,000 years. Hence, the reappearance of the glaciers that form most lakes may occur well before many extant lakes fill in. Thus, conceivably, some lakes will never fill in and will instead survive to the next glacial period, living long enough to be, perhaps, carved out yet again by a glacier. And though glaciers can destroy extant lakes by depositing moraine within them, one can still imagine a lake that is effectively immortal, continuously saved from death by the arrival of a sediment-gouging glacier.

But these days, what is far more of a threat to lakes is not the natural process of sedimentation, but instead various destructive human activities. We have caused, directly or indirectly, very rapid transformations of many lakes, causing them to disappear, sometimes with alarming rapidity and with environmentally disastrous consequences. This is a story that is true for a great many lakes, but is perhaps best exemplified by central Asia's Aral Sea, an area within Kazakhstan and Uzbekistan, a lake that is now almost gone.

Once, the Aral Sea was the fourth largest lake in the world by area, its surface extending over a mammoth 26,064 square miles. Its salinity was about 10 ppt, about one-third that of the ocean. Enormous quantities of fish called the Aral Sea home, and a robust fishing industry existed along its shores. But, starting in the 1960s, withdrawals of freshwater from the Aral's main rivers ramped up significantly, primarily to provide irrigation for expanding agricultural activities. This reduced the inflow to the Aral Sea from about 13 cubic miles per year in 1960 to about 2 cubic miles per year in 2001—a dramatic decrease. In 1960 the Aral Sea had a volume of 261 cubic miles. As of 2006, it was down to 25.9 cubic miles, one-tenth its original size. In the process, the lake was split into two bodies of water, the declining water level having exposed a shallow portion of the lake floor that separates the lake into two basins. The lake surface area decreased from 26,064 square miles in 1960 to 6712 square miles in 2006, and to a paltry 4684 square miles in 2011, a devastating reduction of 21,380 square miles of lake area. And, over this same period, the salinity increased by more than a factor of ten.

The consequences of the numbers cited above were severe for the millions of people who lived in the Aral Sea region. By the 1980s, the fishing industry had collapsed as the fish perished from the increased salinity. Tens of thousands of people were put out of work and at least one species of fish became extinct. Thirty-three species of birds that used to thrive in the wetlands along the shore have simply disappeared. The loss of lake surface area was, of course, matched by the growth of dry lakebed, a gigantic swath of dry, salty earth. Shorelines receded enormous distances, close to 100 miles in the region of the southeastern shore, a change easily visible from satellite. The effect of the desiccation was like adding two Massachusetts' worth of dry, dusty land to the nations of Kazakhstan and Uzbekistan. News reports and magazine stories about this region are rife with images of large, rusted fishing vessels lying on their side in the middle of what is now a desert, the lake having vacated their location years earlier, not a drop of water in sight.

The exposed salty earth has had deleterious impacts on the health of those living in the Aral Sea region. Winds often create large dust storms that sweep huge clouds of dried, salty lake sediment into the air. Plumes of dust have been observed via satellite extending over 300 miles downwind. High levels of respiratory illnesses have been reported, along with increased rates of esophageal cancer and eye maladies. The problem of the dust storms is exacerbated by the fact that as the lake receded, pollutants such as agricultural pesticides and herbicides that had washed into the Sea over the years were concentrated by evaporation and ultimately deposited on the dry lake bottom. Dust storms kick these pollutants into the air, contributing to health problems throughout the region.

And fears of even deadlier problems exist. When the region was part of the Soviet Union, the biological weapons test site, Aralsk-7, was located on what was then an island, Vozrozhdeniya Island. Here, genetically modified smallpox, anthrax, typhus, and other deadly diseases were tested. Although the Soviets claimed to have decontaminated the island, and while the United States sent experts to help Uzbekistan ensure the destruction of any surviving bioagents, there is always the fear that some linger. Vozrozhdeniya Island is no longer an island, but simply part of the shore of the dramatically reduced Aral Sea where any insect, reptile, or mammal is free to enter and depart as it sees fit. As of this writing, the Wikipedia entry for the Aral Sea begins, "The Aral Sea was...". The use of the past tense, though formally incorrect, soberly captures the truth as no statistic or factual statement truly can.

Although the Aral Sea disaster is certainly the most extreme example of modern salt-lake desiccation, it is definitely not the only case. Desiccation is occurring everywhere, from the Great Salt Lake in the United States to Lake Urmia in northwestern Iran, Lop Nur Lake in northwestern China, and Lake Corangamite in Victoria, Australia. In California, Owens Lake was completely dry in 1940, the result of diversions of freshwater to Los Angeles, resulting in unsafe dust levels in the area that are thought to have increased the prevalence of asthma and other respiratory problems in the region. Not far from Yosemite National Park in eastern California, the desiccation of Mono Lake has resulted in toxic alkali dust storms over the exposed salt flats. This drying up of salt lakes all over the world threatens humans, economies, bird habitats, and water creatures such as brine shrimp. Without concerted human efforts, these lakes may disappear leaving little but salt flats where once there were lakes.

Salt lakes are certainly not the only kind of lakes under threat. Freshwater lakes are also increasingly oversubscribed in terms of the amount of water withdrawn from them or from the rivers that feed them. Lake Chad, which serves as a water source for tens of millions of people in West Africa, had a surface area of 9650 square miles in the 1960s and has shrunk to an area of less than 965 square miles, a consequence of irrigation and population growth. Threats like these exist for both naturally occurring lakes and manmade reservoirs. Lake Lanier, a reservoir built in the 1950s that serves as the primary water source for Atlanta and its suburbs, has become oversubscribed as a consequence of the population growth of the nation's ninth-largest metropolitan area. Indeed, during the 2007 drought, the surface of Lake Lanier dropped to the point where Atlanta had less than 90 days of water supply remaining. Although water would still have been left in Lake Lanier, the water level would have fallen below the inlet to the city's water treatment plants, leaving faucets dry, a situation that was narrowly averted by rains that finally alleviated the drought.

In the arid West, the worries are constant. The city of Las Vegas, which gets 90 percent of its water from Lake Mead, contended with similar problems. Faced with a combination of population growth and a significant drought, Las Vegas was confronted with the possibility of a water level that would fall beneath the city's inlet. At "full pond," the surface of Lake Mead is at 1221 feet above sea level (a level it attained only twice, first in 1941 shortly after the Hoover Dam was completed and then again in 1983). Las Vegas had access to two inlets, one at 1050 feet and another at 1000 feet. But typical water levels for Lake Mead are far below full pond and periodically flirt with going lower than both of the Las Vegas inlets. At the time of this writing, Lake Mead is at about 1090 feet and has been dropping more or less steadily since 2000. To deal with the continual water shortage threats, the Southern Nevada Water

Authority created a third water inlet in Lake Mead in 2015. Coined "The Third Straw," the inlet is at an elevation of 860 feet and was built at a cost of $817 million. One hopes this will satisfy the needs of Las Vegas far into the future, though the continuing growth of this part of the nation, and the ongoing droughts in the Southwest leave little room for optimism.

Even the Great Lakes are not immune to humanity's ever-growing thirst. One might think this enormous lake system is safe given that it contains about 20 percent of the world's surface freshwater and is surrounded by cities whose populations tend to be in decline. But having an enormous amount of water at your disposal does not mean that this water cannot be completely taken away. As Dan Eagen argues in his 2017 book *The Death and Life of the Great Lakes,* just look at the Aral Sea. In the course of half a century, it went from the fourth-largest lake in the world to something like 10 percent of its former self, solely due to withdrawals for irrigation and other human uses. In trying to communicate just how much water was removed from the Aral Sea, Eagan quotes *Newsweek* correspondent Peter Annin who travelled to the Aral Sea:

> *Trying to describe what the Aral Sea is like is one of the most frustrating exercises of my journalism career. When you drive for five hours on the old seabed in a Russian jeep from the old shoreline to the new shoreline, how do you quantify that to someone who has never been there? How do you describe the magnitude of the problem when you stop and get out and look around in all directions of the compass and you can't see water anywhere and you know it was once 45 to 50 feet deep over your head?*

Clearly, the size of the Great Lakes cannot be used as an argument against the possibility of their demise.

But to date, the story of the Great Lakes is a good one. A compact currently exists among the states and provinces bordering them prohibiting water withdrawals by counties that aren't located within the Great Lakes basin. This ensures that, one way or another, water withdrawn from these lakes is ultimately returned to it. This is a smart idea since the water used within the Great Lakes basin naturally drains back to those lakes. If you are washing your car in Chicago, the water will come from Lake Michigan, but it will run down a storm drain back into that lake. If you live in Cleveland, your toilet tank was filled by water from Lake Erie, and that's where the water will return once the toilet is flushed, having first been processed at a wastewater treatment plant. And, if you irrigate your garden or lawn somewhere in the suburbs of Toronto, your water will have originated in Lake Ontario, to which any runoff will return. The compact signed in 2008 even has a contingency for counties that are only partially inside the Great Lakes watershed. In this case, municipalities within that county, but outside the watershed can request permission to pipe Great Lakes water in, but only if they agree to treat and return the wastewater back to the watershed. Indeed, in 2016, the Milwaukee suburb of Waukesha was the first entity to obtain such permission, though not without significant protest and controversy.

But though the Great Lakes Compact seems to be functioning well, there have been many attempts in the past to get at Great Lakes water, and these bids continue. Referring again to Eagan's *The Death and Life of the Great Lakes*, proposals have been put forth to pump water from the Great Lakes to the Missouri River, which would in turn be used to recharge the declining Ogallala aquifer, the aquifer underneath many of the High Plains states that supports the American "breadbasket" via irrigation. A businessman once obtained a permit to move Lake Superior water via tanker in order to provide clean drinking water to poor Asian nations. New York City has proposed using Great Lakes water to replenish its reservoirs during drought years. Furthermore, despite protests, bottled water producers have obtained permission to sell Great Lakes water as evidenced by Nestle which, as of 2018, is permitted to pump and bottle 576,000 gallons of water per day from its White Pine Springs well, which is within the Great Lakes basin. Although the Nestle example involves a relatively small volume of water compared to the more egregious proposals cited above, it shows how difficult it is to keep water from being moved and how difficult it is to secure guarantees that a lake or set of lakes will be protected.

And it isn't just the elimination of water due to the needs of thirsty cities and municipalities that are a threat to the world's lakes. Biological threats exist as well. As soon as connections are made between watersheds, species are able to hop between them. Combine this with the tendency we humans have to move plants, fish, and mammals around the world either unintentionally, or to serve as pets, game, or garden trophies, and a recipe for ecological disaster can exist.

Zebra mussels are a major concern. They first invaded the Great Lakes in the 1980s, likely in the ballast tanks of shipping vessels, and they have thoroughly colonized these lakes. A female can lay one million eggs and when young, the offspring are transported by lake currents, enabling them to quickly spread throughout lake systems. Zebra mussel densities can exceed 500,000 mussels per square yard and they are very effective filterers, making the water very clear, but starving indigenous species in the process.

Currently the Great Lakes are also threatened by Asian carp. Bighead carp, grass carp, black carp, and silver carp (collectively referred to as Asian carp) were introduced in the United States in the 1970s as a way to control weeds and algae in canals. These fish are extremely invasive and since escaping where they were originally introduced, they have dominated the rivers they've migrated to and seriously damaged native fish populations. Among these are the Ohio, Missouri, Mississippi, and Illinois rivers. The last of these is especially concerning because there is a direct route to Lake Michigan from the Illinois River via the Chicago Sanitary and Ship Canal. Extraordinary efforts have been implemented to prevent Asian carp from entering Lake Michigan. These methods have included the positioning of high-voltage plates in locks, poisoning of specific portions of the canal, and more. So far these efforts have succeeded, but given how few fish it would take to colonize the lake, it is difficult to be optimistic these approaches will prevail.

If history has taught us anything about lakes it is that they are fragile. They seem immense to us. The enormity of a lake that takes days to cross by boat, tempts us to think that they are invincible—that they are big, and we are small. But history tells us that as big as lakes are, they are not very big. We must recognize that our actions, our population growth, our agricultural methods, all of these are what is big. In comparison to human thirst, even the greatest of lakes are but puddles.

14. TECTONICS

Admittedly, investigating all of the different ways a lake can die can be a bit dreary. Perhaps a more optimistic way to approach the subject is to consider how long a lake can live. In a certain sense we have already addressed that question earlier in this book when we noted Lake Baikal is thought to be the oldest lake in the world with an estimated age of 25 million years. But much like the study of longevity in humans, what is really interesting in the study of longevity of lakes isn't the simple statement of who or what is the oldest (in the case of humans it is Jeanne Louise Calment who is thought to be the oldest human to ever live, dying at the age of 122), but rather to determine what the upper limit to a lifetime might be. Much as scientists seek to determine what exactly keeps a human from living for, say, a thousand years, here we shall try to discern what prevents lakes from living far longer than old Lake Baikal. Is it possible for a lake to exist for a hundred million years? A billion years?

Most lakes will disappear via the processes described in earlier chapters: by filling in with sediment or drying out due to river diversions. Future glaciation may kill some lakes even as it forms or modifies others. There is the ever-changing climate, which by itself may dry out a lake (though the basin will remain). At the edges of continents, lakes may be, and have been, submerged by rising seas. But if some very sturdy lake was able to escape death by these processes, what is the ultimate limit in the age of that lake?

One way to answer this question is to focus on the lake basin itself. Without a basin—the bowl-shaped depression containing a lake—one cannot have a lake. So, if by luck or location a lake is not dried out nor submerged nor filled in, the only thing that could destroy it would have to be a process that destroys the very land it exists upon. One thing that could do this and which would seem to serve as an absolute upper bound on the age of a lake would be a force that destroys the tectonic plate upon which the lake resides. Then, inarguably, a lake would cease to exist. But what destroys tectonic plates? What forces eliminate the very land upon which you and I now reside? To explore this question, let's take a brief look at plate tectonics.

● ● ●

Our planet's shell, the lithosphere, consists of plates, massive expanses of the Earth's crust that rest upon a sea of magma. This is the solid surface that supports the oceans and the mountains; it is where you and I reside. Tectonic plates provide the ground upon which rivers flow and, of course, upon which lakes reside. Though tectonic plates move very little over the time scale of our lives, traveling on average less than a few inches per year, over the time scale of Earth's existence, they have moved quite a bit.

Indeed, the geography of this planet has undergone dramatic changes. If we were able to travel a few hundred million years back in time, we would walk upon an Earth where what is now Africa, North America, South America, and other continents, were all one supercontinent called Pangea. We would find the modern continents that Pangea comprised are gathered together, each located far from where they are found today. This Earth would seem very strange to us, this place where the Atlantic Ocean had not yet opened up—a place where one could hypothetically walk from India to Antarctica. This ancient world would be unrecognizable to us, this place with oceans very different from those that exist today, oceans that geologists have named the Tethys Sea and the Iapetus Ocean. This ancient world would be a place where what is now land was sea and what is now sea was land.

But what *would* look familiar to us would be the shapes of the tectonic plates—the general outline of the continents. South America, though rotated about and located far from its current location, would look similar to its current shape. On a map of the world from this ancient time, you would easily recognize the shape of Africa. Though the locations of continents do change quite a bit over geological time scales, the continents themselves are fairly robust.

Because the planet is completely covered by tectonic plates, their motion causes quite a bit of plate-to-plate interaction. This interaction between plates occurs in three basic ways. Plates can collide with each other, forming convergent boundaries; plates can spread apart from each other creating divergent boundaries; plates can slide past each other. The first of these, plate collision, can result in subduction, the destructive process whereby one plate is pressed beneath another, the subducted plate forced down into the magma beneath, destroying all its natural history in the process. This presents a possible mechanism by which a lake could be truly destroyed. This process tends to occur where an oceanic plate collides with a continental plate. But the oceanic plate is the one that subducts; this is because the density of oceanic plates tends to be much greater than that of continental plates. Thus, what would be destroyed by subduction would not be a lake, but at most a relict lake, a lake that once existed but was flooded by an ocean millions of years earlier.

When tectonic plates separate from each other to form divergent plate boundaries, the tendency is to actually form lakes, not destroy them. The faults that form at divergent plate boundaries sometimes occur as two parallel lines or cracks. The crust laying between these two cracks tends to subside, creating large, often flat depressions. When these depressions fill with water, lakes called graben lakes form. Unlike glacially formed lakes, which were formed during some part of the last glaciation and are therefore on the order of only 100,000 years old, graben lakes can be millions of years old. Lake Baikal is an example of a graben lake. Other examples include Lake Tahoe in North America, several lakes in eastern Africa along the Great Rift Valley, including Lakes Albert, Tanganyika, Edward, Malawi and Turkana (some of the largest lakes in all of Africa), and others.

Not all graben lakes are old. One of the most interesting graben lakes, and one that is not very old, is Lake Thingvallavatn in Iceland. This impressive body of water is thought to have been around for a mere 10,000 years—a veritable youngster. Rifting occurs throughout the Mid-Atlantic Ridge, which runs from south of the equator to beyond the Arctic Circle. Along it, the North American and European continental plates are separating from each other. For most of its length, this ridge runs along the bottom of the Atlantic Ocean, but along its northern portion it runs through Iceland, where it has formed many lakes. Thingvallavatn is one such lake and is celebrated in limnology textbooks. For, although a tectonically formed lake and therefore one we would expect to be quite old, the rifting that formed it is relatively recent.

Also, Thingvallavatn is an example of a lake being formed by multiple processes all at once. Though initially formed by tectonic rifting, it has also been exposed to glaciation, which has eroded and modified its basin. Additionally, lava flows have entered the lake four times in its relatively short history, further modifying the lake morphology. Thingvallavatn is also unique in that it is one of the few places where you can actually see the European and North American tectonic plates separating (at least without descending to the bottom of the Atlantic Ocean), a fact pointed out by many Icelandic travel agencies and travel websites. To further add to its allure, it is along the shores of Thingvallavatn that Iceland's parliament, the Althing, was formed in AD 930. The Althing is the oldest continuously functioning parliament in the world;

Thingvallavatn translates as “Lakes of the Fields of Parliament.” Last but not least, this intriguing lake is also beautiful. Its shores are populated by the short, rugged conifers that you tend to find in northern latitudes, with low rocky mountains rising up from both sides of the long shores.

The final way that tectonic plates can move is when they slide past each other. This motion tends to be of the stick-slip type, meaning that stress builds up between the plates, but without motion. Then, when the stress becomes sufficiently high, the plates move suddenly, causing earthquakes. These can cause many changes in the surface of the plates including cracking, but also tilting or subsidence of the surface. Both results form lakes.

The most famous example of a slipping-plate boundary in the United States is California’s San Andreas Fault. But one of the most violent series of earthquakes in North America actually occurred quite far from the West coast, right in the heartland of America. In the early 1800s, at epicenters located near New Madrid, Missouri, three great earthquakes struck, one on December 16, 1811, another on January 23, 1812, and one more on February 7, 1812. These earthquakes and the hundreds of aftershocks that followed served as a defining event for those portions of Missouri, Tennessee, Kentucky, and Arkansas, where complete devastation occurred.

It was an odd time in history. The earthquakes were preceded by a comet that appeared in September, beginning as a small dot in the nighttime sky and growing to become a fearsome thing, a comet with a two-pronged tail and a halo as large as the moon. The tail eventually became large enough to cover half of the sky. Stranger yet, squirrels decided to migrate south at this time and there are accounts of tens of thousands of them on the move. And then there were the earthquakes, a transformative experience for those who lived through them.

Collectively referred to as “the great shakes” or the New Madrid earthquakes, the impact of these earthquakes was enormous. Near New

Madrid, virtually every man-made structure was flattened for miles. Tremors were felt as far away as Montreal and Baltimore. While the seismic equipment necessary to ascertain the precise strength of these earthquakes didn't exist at the time, the USGS notes the region where strong shaking occurred was two to three times larger than that for the 1964 Alaska earthquake, a magnitude 9.2 quake, the largest recorded in North America.

New Madrid is situated on the Mississippi River, and the impact of the quake on the river must have been terrifying to behold. Residents of the region reported intervals where the river flowed backwards. Enormous cracks appeared in the riverbed, draining the river for a period of time and forming temporary waterfalls. Everywhere banks slid into the river, and the river itself was filled with whirlpools, waves, and waterspouts; the water turned blood-red as the clay soil from the riverbed was mixed by the writhing turbulence of the earthquake-stirred water.

As noted above, earthquakes can result in depressions due to local subsidence, which in turn form lakes. Such lakes were formed in abundance in Tennessee, Missouri, and Arkansas during the New Madrid quakes. Perhaps the largest of the resulting lakes is Reelfoot Lake, located near Tiptonville, Tennessee, an interesting lake in its own right. Reelfoot is shallow and marshy, and so studded with stumps and other obstructions that the locals developed the Reelfoot skiff, a boat with an extra link in the oar mechanisms to allow the rower to face the boat's direction of travel and avoid submerged threats.

So, these are the ways in which the motion of tectonic plates forms lakes. But how do they destroy lakes? Or, keeping in line with our original intent to find the maximum possible age of a lake, what does plate tectonics teach us about where an old lake is most likely to survive? As we noted earlier, for certain, that location would not be on an oceanic plate. While depressions exist in oceanic plates, including some that were once above sea level, it is difficult to think of those depressions as lakes given that their time beneath sea level will have killed anything that once resided in them. Moreover, even if we decided to define a depression in an oceanic plate as a lake, oceanic plates are ever in the process of subducting. Moving from an ocean ridge where two plates spread apart from each other, these plates move toward a continental plate where they will almost certainly subduct, the oceanic crust being recycled into the magma beneath. Hence, the continental crust is far older than oceanic crust, having an average age of about 2.4 eons (an eon is one billion years) and with parts of the continental crust as old as 4.0 eons or even older. The oceanic crust, on the other hand is in all cases less than 200 million years in age. To put these numbers in context, Earth is estimated to be 4.54 eons in age meaning that portions of the continents are only slightly younger than Earth itself. Clearly, the continents are the place to look for a very old lake. But which continent? And on what part of that continent?

One approach would be to look for very old rocks—most logically, to look where the oldest rocks ever discovered have been found. This location turns out to be in modern-day northern Quebec. There, along the shores of Hudson Bay, geologists have discovered rocks that are 4.3 eons old. Their age indicates they were formed just after the very first crust was formed on earth, the protocrust. But even if there are rocks this old in Quebec that currently form a basin, this doesn't necessarily mean such a basin represents a 4.3-billion-year-old lake. The rocks found in this part of Quebec are metamorphic, meaning they were composed of sediments that were buried by other sediments and so on until their depth below the surface was of such a magnitude that they were partially or completely melted by the high pressures and temperatures. Because these rocks were below the surface for a very long time, even if they are currently located in a bowl-shaped terrain, this in no way suggests the existence of a lake 4.3 eons old.

Given that a very old lake must have been part of a bowl-shaped topographical structure, perhaps a better approach for determining the maximum age of a lake would be to search for old basins, basins whose topography is naturally stable in some sense. Counterintuitively, one way to do this is to search for old mountains, given that mountain systems have valleys with rivers, along which lakes form. The paternoster lakes described in an earlier chapter are one example. Some mountain systems are quite old, for instance, the Appalachian Mountains are estimated to be close to 500 million years old (half an eon). Of course they were constantly evolving during this time, rising to a height much higher than that of the present day. In fact, the Appalachians were once as high as the present-day Himalayas; constant and considerable weathering and erosion brought them down to their current height. Given all of the erosion—all of that removal of rock and soil—it is unlikely any basin in those mountains (such as a cirque lake) would survive. But maybe it is possible. Perhaps some basin was periodically filled with the products of erosion and weathering but was periodically flushed during flooding. Perhaps that basin remained a basin in spite of the uplifting of the mountains, which could have pushed upward a low basin, turning something concave in shape into something dome shaped. But maybe, just maybe, somewhere amongst the mountain laurel and rhododendron of the present-day Appalachian Mountains lies a pond, one filled with sediment and emptied of sediment, tipped one way and then the other over the millions of years of mountain forming, but without ever completely emptying it of water. Maybe for only 100 million years, maybe longer, but long enough to have secrets at its bottom that are older than the secrets of Lake Baikal.

Of course, 100 million years is far younger than the age of the planet, and only four times older than Lake Baikal. Could lakes survive even longer? It is hard to know. But even if there is a way a lake could live as long as a continental plate, there is still a limit. While oceanic crust subducts under continental crust when plates collide, this does not mean continental crust has everlasting life. When two continental plates collide, one subducts beneath the other. This is the ultimate fate of any part of the lithosphere—to become part of the so-called rock cycle where new rock is formed perhaps as lava coming out of a volcano, creating igneous rock, which in turn is weathered and eroded to form sedimentary rock, which in turn may form metamorphic rock. But all of these rocks at some point will be subducted, returned to the fiery hot magma of the Earth's mantle.

The Book of Job famously states that "The Lord giveth and the Lord taketh away." And so it is for lakes. While there are myriad ways in which lakes are formed, so too are there many ways they can be destroyed. Often it is the simple endless filling of a basin by sediment that eliminates a lake. Or, a lake's boundaries are relentlessly, continuously weathered away. Or desiccation is the agent, either natural or due to human activities. But if you love lakes, take heart. For while it is true that for every lake on the planet, all evidence of its existence will one day be gone, it is also true that this will take quite a while. You will be gone long before your favorite lake disappears. Your favorite lake may survive the repeated cycles of glacial advance and recession. Your favorite lake may move along with its tectonic plate to a different location on the planet. Your favorite lake may live as long as its continent lives. But even if it does, it too will be destroyed by the ultimate end of things on earth—the fiery magma lying beneath us all. The place where you and your lake will one day meet again.

METRIC CONVERSIONS

INCHES	MM
⅛	3.2
⅙	4.2
¼	6.4
⅓	8.5
⅜	9.5
½	13
¾	19
1	25

INCHES	CM
1	2.5
2	5.1
3	7.6
4	10
5	13
6	15
7	18
8	20
9	23
10	25
20	51
30	76
40	102
50	127

FEET	M
1	0.3
2	0.6
3	0.9
4	1.2
5	1.5
6	1.8
7	2.1
8	2.4
9	2.7
10	3
20	6
30	9
40	12
50	15

TEMPERATURES
°C = 5/9 × (°F−32)
°F = (9/5 × °C) + 32

BIBLIOGRAPHY

Introduction

Bromberg, J. P. 1984. *Physical Chemistry*. Boston: Allyn and Bacon, Inc.

Friedman, T. L. August 27, 1986. In Cameroon, scenes of a valley of death. *New York Times*.

Gibbs, W. W. 2001. Out in the cold. *Scientific American*. March 17.

Horne, A. J., and C. R. Goldman. 1994. *Limnology*. New York: McGraw-Hill.

Hu, D. L., B. Chan, and J. W. M. Bush. 2003. The hydrodynamics of water strider locomotion. *Nature*. 424: 663–666.

Krajick, K. 2003. Defusing Africa's killer lakes. *Smithsonian* 34: Issue 6.

Löffler, H. 2004. The Origin of Lake Basins. In P. E. O'Sullivan and C. S. Reynolds, eds. *The Lakes Handbook Volume 1*. Malden, MA: Blackwell Science.

Micklin, P. 2007. The Aral Sea disaster. *Annual Review of Earth and Planetary Sciences* 35: 47–72.

Ross, T. E. 1987. A comprehensive bibliography of the Carolina bays literature. *The Journal of the Elisha Mitchell Scientific Society* 103: 28–42.

Siegert, M. J., J. C. Priscu, I. A. Alekhina, J. L. Wadham, W. G. Lyons. 2016. Antarctic subglacial lake exploration: first results and future plans. *Philosophical Transactions of the Royal Society A* 374: 20140466.

Smolowe, J., B. J. Phillips. 1986. Cameroon the lake of death. *Time* 128: Issue 10.

Tuttle, M. L., M. A. Clark, H. R. Compton, J. D. Devine, W. C. Evans, A. M. Humphrey, G. W. Kling, E. J. Koenigsberg, J. P. Lockwood, G. N. Wagner, 1987. The 21 August 1986 Lake Nyos Gas Disaster, Cameroon. *Open-File Report 87-97*. United States Geological Survey.

Welch, P. S. 1952. *Limnology*. New York: McGraw-Hill Book Company.

Westenburg, C. L., G. A. DeMeo, and D. J. Tanko. 2006. Evaporation from Lake Mead, Arizona and Nevada, 1997–99. *Scientific Investigations Report 2006-5252*. United States Geological Survey.

Wetzel, R. G. 1975. *Limnology*. Philadelphia: W. B. Saunders Company.

Chapter 1

Andrews, J. 1979. The Present Ice Age: Cenozoic. In B. S. John, ed. *The Winters of the World*. New York: John Wiley & Sons. 173–218.

Bennett, M. R. 2003. Ice streams as the arteries of an ice sheet: their mechanics, stability and significance. *Earth-Science Reviews* 61: 309–339.

Burton, H. R. 1981. Chemistry, physics and evolution of Antarctic saline lakes, a review. *Hydrobiologia* 82: 339–362.

Cole, G. A. 1994. *Textbook of Limnology*. Long Grove, Illinois: Waveland Press.

Derbyshire, E. 1979. Glaciers and Environment. In B. S. John, ed. *The Winters of the World*. New York: John Wiley & Sons 58–106.

Gibbs, W. W. 2001. Out in the cold. *Scientific American* March 17.

Golterman, H. L. 1975. *Physiological Limnology: An Approach to the Physiology of Lake Ecosystems*. Amsterdam, The Netherlands: Elsevier Scientific Publishing.

Horne, A. J., and C. R. Goldman. 1994. *Limnology*. New York: McGraw-Hill.

Hutchinson, G. E. 1957. *A Treatise on Limnology, Volume I*. New York: John Wiley & Sons.

Jackson, M. 2019. *The Secret Lives of Glaciers*. Brattleboro, VT: Green Writers Press.

John, B. 1979. Planet Earth and its Seasons of Cold. In B. S. John, ed. *The Winters of the World*. New York: John Wiley & Sons. 9–28.

Löffler, H. 2004. The Origin of Lake Basins. In P. E. O'Sullivan and C. S. Reynolds, eds. *The Lakes Handbook Volume 1*. Malden, MA: Blackwell Science.

Macdougall, D. 2004. *Frozen Earth: The Once and Future Story of Ice Ages*. Berkeley and Los Angeles: University of California Press.

Montañez, I. P., C. J. Poulsen. 2013. The Late Paleozoic ice age: an evolving paradigm. *Annual Review of Earth and Planetary Sciences* 41: 629–656.

Reynolds, C. S. 2004. Lakes, Limnology and Limnetic Ecology: Towards a New Synthesis. In P. E. O'Sullivan and C. S. Reynolds, eds. *The Lakes Handbook Volume 1*. Malden, MA: Blackwell Science.

Russell, I. C. 1897. *Glaciers of North America*. Boston: Ginn Publishers.

Siegert, M. J., J. C. Priscu, I. A. Alekhina, J. L. Wadham, W. G. Lyons. 2016. Antarctic subglacial lake exploration: first results and future plans. *Philosophical Transactions of the Royal Society A* 374: 20140466.

Thorson, R. M. 2009. *Beyond Walden: The Hidden History of America's Kettle Lakes and Ponds*. New York: Walker.

Vincent, W. F. 2018. *Lakes: A Very Short Introduction*. New York: Oxford University Press.

Wetzel, R. G. 1975. *Limnology*. Philadelphia: W. B. Saunders Company.

Woodbury, D. O. 1962. *The Great White Mantle*. New York: The Viking Press, Inc.

Chapter 2

Bressan, D. August 31, 2016. The Killer Lakes Of Africa - A Rare But Dangerous Volcanic Phenomenon. *Forbes*.

Bryson, B. 2003. *A Short History of Nearly Everything*. New York: Broadway Books.

Buckwalter, S. 2005. *Volcanoes: Disaster & Survival*. Berkeley Heights, NJ: Enslow Publishers.

Coker, R. E. 1954. *Streams, Lakes, Ponds*. New York: Harper and Row.

Cole, G. A. 1994. *Textbook of Limnology*. Long Grove, Illinois: Waveland Press.

Decker, R., and B. Decker. 2003. Krakatau: 1883. In N. Harris and L. Armstrong, eds. *Volcanoes*. Farmington Hills, Michigan: Greenhaven Press.

Friedman, T. L. August 27, 1986. In Cameroon, scenes of a valley of death. *New York Times*.

Golterman, H. L. 1975. *Physiological Limnology: An Approach to the Physiology of Lake Ecosystems*. Amsterdam, The Netherlands: Elsevier Scientific Publishing Company.

Harris, N., and L. Armstrong. 2003a. Introduction. In N. Harris and L. Armstrong, eds. *Volcanoes*. Farmington Hills, Michigan: Greenhaven Press.

Harris, N., and L. Armstrong. 2003b. Appendix. In N. Harris and L. Armstrong, eds. *Volcanoes*. Farmington Hills, Michigan: Greenhaven Press.

Holloway, M. February 27, 2001. Trying to tame the roar of deadly lakes. *New York Times*.

Hutchinson, G. E. 1957. *A Treatise on Limnology, Volume I*. New York: John Wiley & Sons.

James, B. September 17, 1992. Recycling a lake's dangerous gas. *New York Times*.

Kling, G. W. 1987. Seasonal mixing and catastrophic degassing in tropical lakes, Cameroon, West Africa. *Science* 237: 1022–1024.

Kling, G. W., M. A. Clark, H. R. Compton, J. D. Devine, W. C. Evans, A. M. Humphrey, E. J. Koenigsberg, J. P. Lockwood, M. L. Tuttle, G. N. Wagner 1987. The 1986 Lake Nyos gas disaster in Cameroon, West Africa. *Science* 236: 169–175.

Krajick, K. 2003. Defusing Africa's killer lakes. *Smithsonian* 34: Issue 6.

Kusakabe, M., G. Z. Tanyileke, S. A. McCord, S. G. Schladow. 2000. Recent pH and CO2 profiles at Lakes Nyos and Monoun, Cameroon: implications for the degassing strategy and its numerical simulation. *Journal of Volcanology and Geothermal Research* 97: 241–260.

Macdonald, G. A. 1972. *Volcanoes*. Englewood Cliffs, New Jersey: Prentice-Hall.

Murck, B. W., B. J. Skinner, and S. C. Porter. 1996. *Environmental Geology*. New York: John Wiley & Sons.

New York Times. January 22, 1991. Warning on gas in Cameroon lake.

Reynolds, C. S. 2004. Lakes, Limnology and Limnetic Ecology: Towards a New Synthesis. In P. E. O'Sullivan and C. S. Reynolds, eds. *The Lakes Handbook Volume 1*. Malden, MA: Blackwell Science.

Schmid, M., M. Busbridge, and A. Wüest. 2010. Double-diffusive convection in Lake Kivu. *Limnology and Oceanography* 55: 225–238.

Shanklin, E. 1988. Beautiful deadly Lake Nyos. *Anthropology Today* 4: 12–14.

Sigurdsson, H., J. D. Devine, F. M. Tchoua, T. S. Presser, M. K. W. Pringle, W. C. Evans. 1987. *Journal of Volcanology and Geothermal Research* 31: 1–16.

Smolowe, J., B. J. Phillips. 1986. Cameroon the lake of death. *Time* 128: Issue 10.

Sullivan, W. January 23, 1987. U.S. and French experts differ on Cameroon gas eruption. *New York Times*.

Tazieff, H. 1989. Mechanisms of the Nyos carbon dioxide disaster and of so-called phreatic steam eruptions. *Journal of Volcanology and Geothermal Research* 39: 109–116.

Tuttle, M. L., M. A. Clark, H. R. Compton, J. D. Devine, W. C. Evans, A. M. Humphrey, G. W. Kling, E. J. Koenigsberg, J. P. Lockwood, G. N. Wagner, 1987. The 21 August 1986 Lake Nyos Gas Disaster, Cameroon. *Open-File Report 87-97*. United States Geological Survey.

United States Forest Service, U. S. Department of Agriculture: https://www.fs.usda.gov/recarea/coconino/recarea/?recid=55122&actid=102.

von Bubnoff, A. 2005. Deadly lakes may explode again. *Nature News* September 26.

Webster, D. 2003. Inside a Volcano. In N. Harris and L. Armstrong, eds. *Volcanoes*. Farmington Hills, Michigan: Greenhaven Press.

Wetzel, R. G. 1975. *Limnology*. Philadelphia: W. B. Saunders Company.

Wilcoxon, K. H. 2003. Vesuvius: Monarch of the Mediterranean. In N. Harris and L. Armstrong, eds. *Volcanoes*. Farmington Hills, Michigan: Greenhaven Press.

Zhang, Y. July 1996. Cracking the killer lakes of Cameroon. *Geoscience News*.

Chapter 3

American Society of Civil Engineering. 2017. *2017 Infrastructure Report Card*.

Chao, B. F., Y. H. Wu, and S. Li. 2008. Impact of artificial reservoir water impoundment on global sea level. *Science* 320: 212–214.

Chen, S., Z. Chen, R. Tao, S. Yu, W. Xu, X. Zhou, Z. Zhou. 2018. Emergency response and back analysis of the failures of earthquake triggered cascade landslide dams on the Mianyuan River, China. *Natural Hazards Review* 19: 05018005.

Cobb, E. S. 2009. Managing Small Fishing Lakes and Ponds in Tennessee. *Tennessee Wildlife Resources Agency* Authorization No. 328838.

Cui, P., Y.-Y. Zhu, Y.-S. Han, X.-Q. Chen, and J.-Q. Zhuang. 2009. The 12 May Wenchuan earthquake-induced landslide lakes: distribution and preliminary risk evaluation. *Landslides* 6: 209–223.

Dai, F. C., C. F. Lee, J. H. Deng, and L. G. Tham. 2005. The 1786 earthquake-triggered landslide dam and subsequent dam-break flood on the Dadu River, southwestern China. *Geomorphology* 65: 205–221.

Downing, J. A., Y. T. Prairie, J. J. Cole, C. M. Duarte, L. J. Tranvik, R. G. Striegl, W. H. McDowell, P. Kortelainen, N. F. Caraco, J. M. Melack, J. J. Middleburg. 2006. The global abundance and size distribution of lakes, ponds, and impoundments. *Limnology and Oceanography* 51: 2388–2397.

Fahlbusch. H. 2009. Early dams. *Engineering History and Heritage* 162: 13–18.

Federal Emergency Management Agency. 2016. *The National Dam Safety Program Biennial Report to the United States Congress.* August 3, 2016.

Helfrich, L. A. 2009. Pond Construction: Some Practical Considerations. *Virginia Cooperative Extension Publication* 420-011.

ICOLD: https://www.icold-cigb.org/

Liu, N., Z. Chen, J. Zhang, W. Lin, W. Chen, and W. Xu. 2010. Draining the Tangjiashan barrier lake. *Journal of Hydraulic Engineering* 136: 914–923.

McDonald, C. P., J. A. Rover, E. G. Stets, and R. G. Striegl. 2012. The regional abundance and size distribution of lakes and reservoirs in the United States and implications for estimates of global lake extent. *Limnology and Oceanography* 57: 597–606.

Messager, M. L., B. Lehner, G. Grill, I. Nedeva, and O. Schmitt. 2016. Estimating the volume and age of water stored in global lakes using a geo-statistical approach. *Nature Communications* 7: 13603.

Morris, J. 2020. Take care of pond quality and fish numbers before winter. *Iowa State University Extension and Outreach News.* https://www.extension.iastate.edu/news/take-care-pond-quality-and-fish-numbers-winter.

Neal, W., D. Riecke, and G. Clardy. 2020. Managing Mississippi Ponds and Small Lakes: A Landowner's Guide. *Extension Service of Mississippi State University, Publication P1428.*

Pokhrel, Y. N., N. Hanasaki, P. J.-F. Yeh, T. J. Yamada, S. Kanae, T. Oki. 2012. Model estimates of sea-level change due to anthropogenic impacts on terrestrial water storage. *Nature Geoscience* 5: 389–392.

Ranganathan, V. 1997. Hydropower and environment in India. *Energy Policy* 25: 435–438.

Reynolds, C. S. 2004. Lakes, Limnology and Limnetic Ecology: Towards a New Synthesis. In P. E. O'Sullivan and C. S. Reynolds, eds. *The Lakes Handbook Volume 1*. Malden, MA: Blackwell Science.

South Carolina Division or Natural Resources: http://dnr.sc.gov/cgi-bin/sco/hsums/cliMAINnew.pl?sc1770.

Tang, C., J. Zhu, X. Qi, J. Ding. 2011. Landslides induced by the Wenchuan earthquake and the subsequent strong rainfall event: A case study in the Beichuan area of China. *Engineering Geology* 122: 22–33.

Taylor, A. 2018. 10 years since the devastating 2008 Sichuan earthquake. *The Atlantic* (May 9).

Wang, B., T. Zhang, Q. Zhou, W. Chao, Y. Chen, W. Ping. 2015. A case study of the Tangjiashan landslide dam-break. *Journal of Hydrodynamics* 27: 223–233.

Zhang, L., M. Peng, D. Chang, and Y. Xu. 2016. *Dam Failure Mechanisms and Risk Assessment*. Singapore: John Wiley & Sons Singapore Pts.

Chapter 4

Allred, K., W. Luoa, M. Konen, B. B. Curry. 2014. Morphometric analysis of ice-walled lake plains in Northern Illinois: Implications of lake elongation by wind-induced dual-cycle currents. *Geomorphology* 220: 50–57.

Alvarez, L. W., W. Alvarez, F. Asaro, and H. V. Michel. 1980. Extraterrestrial cause for the Cretaceous-Tertiary extinction. *Science* 208: 1095–1108.

Ashton, A., A. B. Murray, O. Arnault. 2001. Formation of coastline features by large-scale instabilities induced by high-angle waves. *Nature* 414: 296–300.

Ashton, A. D., A. B. Murray, R. Littlewood, D. A. Lewis, P. Hong. 2009. Fetch-limited self-organization of elongate water bodies. *Geology* 37: 187–190.

Black, R. F., and W. L. Barksdale. 1949. Oriented lakes of Northern Alaska. *The Journal of Geology* 57: 105–118.

Boslough, M. 2012. Inconsistent impact hypotheses for the Younger Dryas. *Proceedings of the National Academy of Sciences* 109: E2241.

Boslough, M. 2013. Faulty protocols yield contaminated samples, unconfirmed results. *Proceedings of the National Academy of Sciences* 110: E1651.

Boslough, M., A. W. Harris, C. Chapman, D. Morrison. 2013. Younger Dryas impact model confuses comet facts, defies airburst physics. *Proceedings of the National Academy of Sciences* 110: E4170.

Boslough, M., K. Nicoll, T. L. Daulton, A. C. Scott, P. Claeys, J. L. Gill, J. R. Marlon, P. J. Bartlein. 2015. Incomplete Bayesian model rejects contradictory radiocarbon data for being contradictory. *Proceedings of the National Academy of Sciences* 112: E6722.

Bunch, T. E., A. West, R. B. Firestone, J. P. Kennett, J. H. Wittke, C. R. Kinzie, W. S. Wolbach. 2010. Geochemical data reported by Paquay et al. do not refute Younger Dryas impact event. *Proceedings of the National Academy of Sciences* 107: E58.

Carlson, A. E. 2010. What caused the Younger Dryas cold event? *Geology* 38: 383–384.

Clark, R. B., T. Poland. 2020. *Carolina Bays*. Columbia, South Carolina: University of South Carolina Press.

Collard, M., B. Buchanan, K. Edinborough. 2008. Reply to Anderson et al., Jones, Kennett and West, Culleton, and Kennett et al.: Further evidence against the extraterrestrial impact hypothesis. *Proceedings of National Academy of Sciences* 105: E112–E114.

Cook, C. W. 2014. Carolina bays: discussion. *Geological Society of America Bulletin* 81: 3171–3172.

Davias, M. E., J. L. Gilbride. 2010. LiDAR imagery employed in Carolina bays research. *Geological Society of America Annual Meeting*. Abstract No. 176738.

Firestone, R. B., A. West, J. P. Kennett, L. Becker, T. E. Bunch, Z. S. Revays, P. H. Schultz, T. Belgya, D. J. Kennett, J. M. Erlandson, O. J. Dickenson, A. C. Goodyear, R. S. Harris, G. A. Howard, J. B. Kloosterman, P. Lechler, P. A. Mayewski, J. Montgomery, R. Poreda, T. Darrah, S. S. Que Hee, A. R. Smith, A. Stich, W. Topping, J. H. Wittke, and W. S. Wolbach. 2007. Evidence for an extraterrestrial impact 12,900 years ago that contributed to the megafaunal extinctions and the Younger Dryas cooling. *Proceedings of the National Academy of Sciences* 104: 16016–16021.

Firestone, R., A. West, and S. Warwick-Smith. 2006. *The Cycle of Cosmic Catastrophes*. Rochester, Vermont: Bear & Co.

Grant, J. A., M. J. Brooks, B. E. Taylor. 1998. New constraints on the evolution of Carolina bays from ground-penetrating radar. *Geomorphology* 22: 325–345.

Hardiman, M., A. C. Scott, M. E. Collinson, R. S. Anderson. 2012. Inconsistent redefining of the carbon spherule "impact" proxy. *Proceedings of the National Academy of Sciences* 109: E2244.

Holliday, V. T., T. Surovell, D. J. Meltzer, D. K. Grayson, and M. Boslough. 2014. The Younger Dryas impact hypothesis: a cosmic catastrophe. *Journal of Quaternary Science* 29: 515–530.

Holliday, V., T. Surovell, E. Johnson. 2016. A blind test of the Younger Dryas impact hypothesis. *PLoS ONE* 11(7): e0155470.

Humphreys, J. 2000. *Nowhere Else on Earth*. New York: Penguin Group.

Israde-Alcántara, I., J. L. Bischoff, G. Dominguez-Vázquez, H.-C. Li, P. S. DeCarli, T. E. Bunch, J. H. Wittke, J. C. Weaver, R. B. Firestone, A. West, J. P. Kennett, C. Mercer, S. Xie, E. K. Richman, C. R. Kinzie, W. S. Wolbach. 2012. Evidence from central Mexico supporting the Younger Dryas extraterrestrial impact hypothesis. *Proceedings of the National Academy of Sciences* 109: E738–747.

Johnson, D. 1937. Role of artesian waters in forming the Carolina bays. *Science* 86: 255–258.

Kaczorowski, R. T. 1977. The Carolina bays: A comparison with modern oriented lakes. Technical Report No. 13-CRD. University of South Carolina, Department of Geology.

Kennett, D. J., J. P. Kennett, G. J. West, J. M. Erlandson, J. R. Johnson, I. L. Hendy, A. West, B. J. Culleton, T. L. Jones, T. W. Stafford, Jr. 2008. Wildfire and abrupt ecosystem disruption on California's Northern Channel Islands at the Ållerød-Younger Dryas boundary (13.0 - 12.9 ka). *Quaternary Science Reviews* 27: 2530–2545.

Kennett, J. P., D. J. Kennett, B. J. Culleton, J. E. A. Tortosa, J. L. Bischoff, T. E. Bunch, I. R. Daniel, Jr., J. M. Erlandson, D. Ferraro, R. B. Firestone, A. C. Goodyear, I. Israde-Alcántara, J. R. Johnson, J. F. J. Pardo, D. R. Kimbel, M. A. LeCompte, N. H. Lopinot, W. C. Mahaney, A. M. T. Moore, C. R. Moore, J. H. Ray, T. W. Stafford Jr., K. B. Tankersley, J. H. Wittke, W. S. Wolbach, A. West. 2015. Bayesian chronological analyses consistent with synchronous age of 12,835–12,735 Cal B. P. for Younger Dryas boundary on four continents. *Proceedings of the National Academy of Sciences* 112: E4344–4353.

Kerr, R. A. 2007. Mammoth-killer impact gets mixed reception from earth scientists. *Science* 316: 1264–1265.

Kerr, R. A. 2008. Experts find no evidence for a mammoth-killer impact. *Science* 319: 1331–1332.

Kerr, R. A. 2009. Did the mammoth slayer leave a diamond calling card? *Science* 323: 26.

Klokočník, J., J. Kostelecký, and A. Bezděk. 2019. The putative Saginaw impact structure, Michigan, Lake Huron, in the light of gravity aspects derived from recent EIGEN 6C4 gravity field model. *Journal of Great Lakes Research* 45: 12–20.

Krajick, K., and C. Davidson. 1997. The riddle of the Carolina bays. *Smithsonian* 28: Issue 6.

LeCompte, M. A., A. C. Goodyear, M. N. Demitroff, D. Batchelor, E. K. Vogel, C. Money, B. N. Rock, A. W. Seidel. 2012. Independent evaluation of conflicting microspherule results from different investigations of the Younger Dryas impact hypothesis. *Proceedings of the National Academy of Sciences* 109: E2960–2969.

LeCompte, M. A., D. Batchelor, M. N. Demitroff, E. K. Vogel, C. Money, B. N. Rock, A. W. Seidel. 2013. Reply to Boslough: Prior studies validating research are ignored. *Proceedings of the National Academy of Sciences* 110: E1652.

Livingstone, D. A. 1954. On the orientation of lake basins. *American Journal of Science* 252: 547–554.

Lyzenga, G. A. 1999. Why are impact craters always round? *Scientific American* October 21.

Mahaney, W. C., V. Kalm, D. H. Krinsley, P. Tricart, S. Schwartz, J. Dohm, K. J. Kim, B. Kapran, M. W. Milner, R. Beukens, S. Boccia, R. G. V. Hancock, K. M. Hart, B. Kelleher. 2010. Evidence from the northwestern Venezuelan Andes for extraterrestrial impact: The black mat enigma. *Geomorphology* 116: 48–57.

May, J. H., A. G. Warne. 1999. Hydrogeologic and geochemical factors required for the development of Carolina bays along the Atlantic and Gulf of Mexico Coastal Plain, USA. *Environmental & Engineering Geoscience* V: 261–270.

Melton, F. A., W. Schriever. 1933. The Carolina "bays": Are they meteorite scars? *Journal of Geology* 41: 52–66.

Meltzer, D. J., V. T. Holliday, M. D. Cannon, D. S. Miller. 2014. Chronological evidence fails to support claim of an isochronous widespread layer of cosmic impact indicators dated to 12,800 years ago. *Proceedings of the National Academy of Sciences* 111: E2162–2171.

Morrison, D. 2010. Did a cosmic impact kill the mammoths? *Skeptical Inquirer* 34: 14–18.

Paquay, F. S., S. Goderis, G. Ravizza, F. Vanhaeck, M. Boyd, T. A. Surovell, V. T. Holliday, C. V. Haynes, Jr., P. Claeys. 2009. Absence of geochemical evidence for an impact event at the Bølling-Allerød/Younger Dryas transition. *Proceedings of the National Academy of Sciences* 106: 21505–21510.

Pinter, N., A. C. Scott, T. L. Daulton, A. Podoll, C. Koeberl, R. S. Anderson, and S. E. Ishman. 2011. The Younger Dryas impact hypothesis: A requiem. *Earth-Science Reviews* 106: 247–264.

Prouty, W. F. 1952. Carolina bays and their origin. *Geological Society of America Bulletin* 63: 167–224.

Rodriguez, A. B., M. N. Waters, M. F. Piehler. 2012. Burning peat and reworking loess contribute to the formation and evolution of a large Carolina-bay basin. *Quaternary Research* 77: 171–181.

Ross, T. E. 1987. A comprehensive bibliography of the Carolina bays literature. *The Journal of the Elisha Mitchell Scientific Society* 103: 28–42.

Schaetzl, R. J., W. Sauck, P. V. Heinrich, P. M. Colgan, and V. T. Holliday. 2019. Commentary on Klokočník, J., J. Kostelecký, and A. Bezděk. 2019. The putative Saginaw impact structure, Michigan, Lake Huron, in the light of gravity aspects derived from recent EIGEN 6C4 gravity field model. Journal of Great Lakes Research 45:12–20. *Journal of Great Lakes Research* 45: 1003–1006.

Schriever, W. 1955. Were the Carolina bays oriented by gyroscopic action? *Science* 121: 806.

Surovell, T. A., V. T. Holliday, J. A. M. Gingerich, C. Ketron, C. V. Haynes, Jr., I. Hilman, D. P. Wagner, E. Johnson, P. Claeys. 2009. An independent evaluation of the Younger Dryas extraterrestrial impact hypothesis. *Proceedings of the National Academy of Sciences* 106: 18155–18158.

Thom, B. G. 1970. Carolina bays in Horry and Marion Counties, South Carolina. *Geological Society of America Bulletin* 81: 783–814.

van Hoesel, A., W. Z. Hoek, G. M. Pennock, M. R. Drury. 2014. The Younger Dryas impact hypothesis: a critical review. *Quaternary Science Reviews* 83: 95–114.

Zamora, A. 2017. A model for the geomorphology of the Carolina Bays. *Geomorphology* 282: 209–216.

Chapter 5

Black, R. F., and W. L. Barksdale. 1949. Oriented lakes of Northern Alaska. *Journal of Geology* 57: 105–118.

Carson, C. E., and K. M. Hussey. 1959. The multiple-working hypothesis as applied to Alaska's oriented lakes. *Proceedings of the Iowa Academy of Science* 66: 334–349.

Carson, C. E., and K. M. Hussey. 1960. Hydrodynamics of three Arctic lakes. *Journal of Geology* 68: 585–600.

Carson, C. E., and K. M. Hussey. 1962. The oriented lakes of Arctic Alaska. *Journal of Geology* 70: 417–439.

Carson, C. E., and K. M. Hussey. 1963. The oriented lakes of Arctic Alaska: A reply. *The Journal of Geology* 71: 532–533.

Carson, C. E. 2001. The oriented thaw lakes: A retrospective. In D. W. Norton, ed. *Fifty More Years Below Zero*. Fairbanks, Alaska: Arctic Institute of North America. 129–138.

Carter, L. D. 1987. Oriented Lakes. In W. L. Graf, ed. *Geomorphic Systems of North America, Vol. 2*. Boulder, Colorado: Geological Society of America. 615–619.

Hopkins, D. M. 1949. Thaw lakes and thaw sinks in the Imuruk Lake area, Seward Peninsula, Alaska. *Journal of Geology* 57: 119–131.

Hutchinson, G. E. 1957. *A Treatise on Limnology, Volume I*. New York: John Wiley & Sons.

Ji, Z.-G., K.-R. Jin. 2006. Gyres and seiches in a large and shallow lake. *Journal of Great Lakes Research* 32: 764–775.

Livingstone, D. A. 1954. On the orientation of lake basins. *American Journal of Science* 252: 547–554.

Livingstone, D. A., K. Bryan, Jr., R. G. Leahy. 1958. Effects of an Arctic environment on the origin and development of freshwater lakes. *Limnology and Oceanography* 3: 192–214.

Price, W. A. 1963. The oriented lakes of Arctic Alaska: A discussion. *Journal of Geology* 71: 530–531.

Rosenfeld, G. A., K. M. Hussey. 1958. A consideration of the problem of oriented lakes. *Proceedings of the Iowa Academy of Science* 65: 279–287.

Rex, R. W. 1961. Hydrodynamic analysis of circulation and orientation of lakes in northern Alaska. In G. O. Raasch, ed. *First International Symposium on Arctic Geology Proceedings*. Toronto: Toronto University Press. 1021–1043.

Uda, T., M. Serizawa, T. San-nami, S. Miyahara. 2014. Prediction of formation of oriented lakes. *Coastal Engineering Proceedings*. No. 34.

Zhan, S., R. A. Beck, K. M. Hinkel, H. Liu, and B. M. Jones. 2014. Spatio-temporal analysis of gyres in oriented lakes on the Arctic Coastal Plain of Northern Alaska based on remotely sensed images. *Remote Sensing* 6: 9170–9193.

Chapter 6

Bell, R. E., M. Studinger, A. A. Tikku, G. K. C. Clarke, M. M. Gutner, C. Meertens. 2002. Origin and fate of Lake Vostok water frozen to the base of the East Antarctic ice sheet. *Nature* 416: 307–310.

Bulat, S. A. 2016 Microbiology of the subglacial Lake Vostok: first results of borehole-frozen lake water analysis and prospects for searching for lake inhabitants. *Philosophical Transactions of the Royal Society A* 374: 20140292.

Christner, B. C., E. Mosley-Thompson, L. G. Thompson, J. N. Reeve. 2001. Isolation of bacteria and 16S rDNAs from Lake Vostok accretion ice. *Environmental Microbiology* 3: 570–577.

Christner, B. C., G. Royston-Bishop, C. M. Foreman, B. R. Arnold, M. Tranter, K. A. Welch, W. B. Lyons, A. I. Tsapin, M. Studinger, J. C. Priscu. Limnological conditions in Subglacial Lake Vostok, Antarctica. *Limnology and Oceanography* 51: 2485–2501.

Gibbs, W. W. 2001. Out in the cold. *Scientific American* March 17.

Gramling, C. 2012. A tiny window opens into Lake Vostok, while a vast continent awaits. *Science* 335: 788–789.

Gura, C., S. O. Rogers. 2020. Metatranscriptomic and metagenomic analysis of biological diversity in subglacial Lake Vostok (Antarctica). *Biology* 9: doi:10.3390.

Inman, M. 2005. The plan to unlock Lake Vostok. *Science* 310: 611–612.

Karl, D. M., D. F. Bird, K. Björkman, T. Houlihan, R. Shackelford, L. Tupas. 1999. Microorganisms in the accreted ice of Lake Vostok, Antarctica. *Science* 286: 2144–2147.

Lukin, V. V, N. I. Vasiliev. 2014. Technological aspects of the final phase of drilling borehole 5G and unsealing Vostok subglacial lake, East Antarctica. *Annals of Glaciology* 55: 83–89.

National Research Council, Committee on Principles of Environmental Stewardship for the Exploration and Study of Subglacial Environments. 2007. *Exploration of Antarctic Subglacial Aquatic Environments*. National Academy of Sciences.

Orosei, R., S. E. Lauro, E. Pettinelli, A. Cicchetti, M. Coradini, B. Cosciotti, F. Di Paolo, E. Flamini, E. Mattei, M. Pajola, F. Soldovieri, M. Cartacci, F. Cassenti, A. Frigeri, S. Giuppi, R. Martufi, A. Masdea, G. Mitri, C. Nenna, R. Noschese, M. Restano, and R. Seu. 2018. Radar evidence of subglacial liquid water on Mars. *Science* 361: 490–493.

Pattyn, F., S. P. Carter, M. Thoma. 2016. Advances in modelling subglacial lakes and their interaction with the Antarctic ice sheet. *Philosophical Transactions of the Royal Society* A 374: 20140296.

Reed, C. February 2000. Bacteria in Lake Vostok. *Geotimes*.

Richter, A., S. V. Popov, L. Schröder, J. Schwabe, H. Ewert, M. Scheinert, M. Horwath, and R. Dietrich. 2014. Subglacial Lake Vostok not expected to discharge water. *Geophysical Research Letters* 41: 6772–6778.

Rutishauser, A., D. D. Blankenship, M. Sharp, M. L. Skidmore, J. S. Greenbaum, C. Grima, D. M. Schroeder, J. A. Dowdeswell, and D. A. Young. 2018. Discovery of a hypersaline subglacial lake complex beneath Devon Ice Cap, Canadian Arctic. *Science Advances* 4: doi:10.1126.

Sarkar, S. 2012. Drilling at Lake Vostok by the Russians. *Current Science* 102: 1355.

Sherstyankin, P. P., L. N. Kuimova, V. L. Potemkin. 2014. Thermodynamic parameters of water in subglacial Lake Vostok, Eastern Antarctica. *Doklady Earth Sciences* 454: 163–168.

Shreve, R. L. 1972. Movement of water in glaciers. *Journal of Glaciology* 11: 205–214.

Siegert, M. J. 1999. Antarctica's Lake Vostok. *American Scientist* 87: 6.

Siegert, M. J., J. C. Ellis-Evans, M. Tranter, C. Mayer, J.-R. Petit, A. Salamatin, J. C. Priscu. 2001. Physical, chemical and biological processes in Lake Vostok and other Antarctic subglacial lakes. *Nature* 414: 603–609.

Siegert, M. J. 2005. Lakes beneath the ice sheet: the occurrence, analysis, and future exploration of Lake Vostok and other Antarctic subglacial lakes. *Annual Review of Earth and Planetary Sciences* 33: 215–245.

Siegert, M. J., K. Makinson, D. Blake, M. Mowlem, and N. Ross. 2014. An assessment of deep hot-water drilling as a means to undertake direct measurement and sampling of Antarctic subglacial lakes: experience and lessons learned from the Lake Ellsworth field season 2012/13. *Annals of Glaciology* 55: 59–73.

Siegert, M. J. 2018. A 60-year international history of Antarctic subglacial lake exploration. In M. J. Siegert, S. S. R. Jamieson, and D. A. White, eds. *Exploration of Subsurface Antarctica: Uncovering Past Changes and Modern Processes*. London: U.K. Geological Society.

Sinha, R. K., K. P. Krishnan. 2013. Novel opportunity for understanding origin and evolution of life: perspectives on the exploration of subglacial environment of Lake Vostok, Antarctica. *Annals of Microbiology* 63: 409–415.

Stone, R. 1999. Lake Vostok probe faces delays. *Science* 286: 36–37.

Stone, R. 2000. Vostok: looking for life beneath an Antarctic glacier. *Smithsonian* 31: 92–100.

Watson, E. 2003. Lake Vostok is like a giant can of soda. *New Scientist* 179: 21.

Wells, M. G., J. S. Wettlaufer. 2008. Circulation in Lake Vostok: A laboratory analogue study. *Geophysical Research Letters* 35: L03501.

Witze, A. 2018. Signs of buried lake on Mars tantalize scientists. *Nature* 560: 13–14.

Wright, A., M. J. Siegert. 2011. The identification and physiographical setting of Antarctic subglacial lakes: An update based on recent discoveries. In M. J. Siegert and M. C. Kennicutt, II, eds. *Antarctic Subglacial Environments*. Washington, D.C.: American Geophysical Union.

Wüest, A., and E. Carmack. 2000. A priori estimates of mixing and circulation in the hard-to-reach water body of Lake Vostok. *Ocean Modelling* 2: 29–43.

Chapter 7

Boetius, A., S. Joye. 2009. Thriving in salt. *Science* 324: 1523–1525.

Cassela, C. December 14, 2016. How an Australian lake turned bubblegum pink. *Australian Geographic*.

Cole, G. A. 1994. *Textbook of Limnology*. Long Grove, Illinois: Waveland Press.

Gavrieli, I., A. Bein, A. Oren. 2005. The expected impact of the peace conduit project (the Red Sea - Dead Sea pipeline) on the Dead Sea. *Mitigation and Adaptation Strategies for Global Change*. 10: 3–22.

Ghazleh, S. A., A. M. Abed, S. Kempe. 2011. The dramatic drop of the Dead Sea: background, rates, impacts and solutions. In V. Badescu and R. B Cathcart eds. *Macro-engineering Seawater in Unique Environments*. Berlin. Springer-Verlag. 77–105.

Hammer, U. T. 1986. Saline lake resources of the Canadian Prairies. *Canadian Water Resources Journal* 11: 43–57.

Horne, A. J., and C. R. Goldman. 1994. *Limnology*. New York: McGraw-Hill.

Hutchinson, G. E. 1957. *A Treatise on Limnology, Volume I*. New York: John Wiley & Sons.

Kalinin, G. B., V. D. Bykov. 1969. The world's water resources, present and future. *Impact of Science on Society* XIX: 135–150.

Messager, M. L., B. Lehner, G. Grill, I. Nedeva, and O. Schmitt. 2016. Estimating the volume and age of water stored in global lakes using a geo-statistical approach. *Nature Communications* 7: 13603.

Micklin, P. 2007. The Aral Sea disaster. *Annual Review of Earth and Planetary Sciences* 35: 47–72.

Ouillon, R., N. G. Lensky, V. Lyakhovsky, A. Arnon, E. Meiburg. 2019. Halite precipitation from double-diffusive salt fingers in the Dead Sea: numerical simulations. *Water Resources Research* 55: 4252–4265.

Patowary, K. November 5, 2018. Brine pools: the underwater lakes of despair. *Amusing Planet*. https://www.amusingplanet.com/2018/11/brine-pools-lakes-under-ocean.html.

Sirota, I., A. Arnon, N. G. Lensky. 2016. Seasonal variations of halite saturation in the Dead Sea. *Water Resources Research* 52: 7151–7162.

Vallentyne, J. R. 1972. Freshwater supplies and pollution: effects of the demophoric explosion on water and man. In N. Polunin, ed. *The Environmental Future*. London, U.K. Macmillan Press. 181–199.

Warren, J. K. 2006. *Evaporites: Sediments, Resources and Hydrocarbons*. Berlin, Germany. Springer-Verlag.

Welch, P. S. 1952. *Limnology*. New York: McGraw-Hill Book Company.

Wetzel, R. G. 1975. *Limnology*. Philadelphia: W. B. Saunders Company.

Whitehurst, L. 2020. Rare salt formations appear along the Great Salt Lake. *Phys.org*.

Williams, W. D. 1996. The largest, highest, and lowest lakes of the world: Saline lakes. *Verhandlungen der Internationalen Vereinigung für Theoretische und Angewandte Limnologie* 26: 61–79.

Wurtsbaugh, W. A., C. Miller, S. E. Null, R. J. DeRose, P. Wilcock, M. Hahnenberger, F. Howe, J. Moore. 2017. Decline of the world's saline lakes. *Nature Geoscience* 10: 816–821.

Chapter 8

Ashby, S. L., R. H. Kennedy, J. H. Carroll, and J. J. Hains. 1994. Water quality studies: Richard B. Russell and J. Strom Thurmond Lakes; summary report. Miscellaneous Paper EL-94-6. U.S. Army Corps of Engineers, Vicksburg, Mississippi.

Cole, G. A. 1994. *Textbook of Limnology*. Long Grove, Illinois: Waveland Press.

Gavrieli, I., A. Bein, A. Oren. 2005. The expected impact of the peace conduit project (the Red Sea - Dead Sea pipeline) on the Dead Sea. *Mitigation and Adaptation Strategies for Global Change* 10: 3–22.

Golterman, H. L. 1975. *Physiological Limnology: An Approach to the Physiology of Lake Ecosystems*. Amsterdam, The Netherlands: Elsevier Scientific Publishing Company.

Holland, J. P., and C. H. Tate. 1984. Investigation and discussion of techniques for hypolimnion aeration/oxygenation. Technical Report E-84-10. U.S. Army Corps of Engineers, Vicksburg, Mississippi.

Horne, A. J., and C. R. Goldman. 1994. *Limnology*. New York: McGraw-Hill.

Lemons, J. W., M. C. Vorwerk, and J. H. Carroll. 1998. Determination of Richard B. Russell dissolved oxygen injection system efficiency utilizing automated remote monitoring technologies. U.S. Army Corps of Engineers, Vicksburg, Mississippi.

Mercier, P., and J. Perret. 1949. Aeration station of Lake Bret, Monastbull, Schwiez. *Ver. Gas. Wasser-Fachm* 29: 25.

Pankow, J. F. 1991. *Aquatic Chemistry Concepts*. Boca Raton, FL: CRC Press.

Ruby, R. 1995. *Jericho: Dreams, Ruins, Phantoms*. New York: Henry Holt.

Singleton, V. L., J. C. Little. 2006. Designing hypolimnetic aeration and oxygenation systems—a review. *Environmental Science and Technology* 40: 7512–7520.

Vincent, W. F. 2018. *Lakes: A Very Short Introduction*. New York: Oxford University Press.

Warren, J. K. 2006. *Evaporites: Sediments, Resources and Hydrocarbons*. Berlin, Germany: Springer-Verlag.

Welch, P. S. 1952. *Limnology*. New York: McGraw-Hill Book Company.

Chapter 9

Allan, J. D. 1997. *Stream Ecology*. London, U.K.: Chapman & Hall.

Cole, G. A. 1994. *Textbook of Limnology*. Long Grove, Illinois: Waveland Press.

Horne, A. J., and C. R. Goldman. 1994. *Limnology*. New York: McGraw-Hill.

Wetzel, R. G. 1975. *Limnology*. Philadelphia: W. B. Saunders Company.

Chapter 10

Bass, R. 1998. The Hermit's Story. *The Paris Review*. 147.

Bass, R. 1999. The Hermit's Story. In A. Tan, K. Kenison, eds. *The Best American Short Stories 1999*. New York: Houghton Mifflin Company.

Bromberg, J. P. 1984. *Physical Chemistry*. Boston: Allyn and Bacon, Inc.

Glasstone, S. 1948. *Elements of Physical Chemistry*. New York: D. Van Nostrand Company, Inc.

Vonnegut, K. 1963. *Cat's Cradle*. New York: Holt, Rinehart & Winston.

Weast, R. C., ed. 1971. *Handbook of Chemistry and Physics, 52nd Edition*. Cleveland, Ohio: Chemical Rubber Company.

Chapter 11

Bower, S. M., and J. R. Saylor. 2013. Sherwood-Rayleigh parameterization for evaporation in the presence of surfactant monolayers, *American Institute of Chemical Engineering Journal* 59: 303–315.

Bower, S. M., and J. R. Saylor. 2011. The effects of surfactant monolayers on free surface natural convection, *International Journal of Heat and Mass Transfer* 54: 5348–5358.

Bush, J. W. M., and D. L. Hu. 2006. Walking on water: Biolocomotion at the interface. *Annual Review of Fluid Mechanics* 38: 339–369.

Clavijo, C. E., J. Crockett, D. Maynes. 2016. Wenzel to Cassie transition during droplet impingement on a superhydrophobic surface. *Physical Review Fluids* 1: 073902.

Erbil, H. Y. 2020. Practical applications of superhydrophobic materials and coatings: Problems and perspectives. *Langmuir* 36: 2493–2509.

Gao, X., and L. Jiang. 2004. Water-repellent legs of water striders. *Nature* 432: 36.

Guo, Z., and W. Liu. 2007. Biomimic from the superhydrophobic plant leaves in nature: Binary structure and unitary structure. *Plant Science* 172: 1103–1112.

Hu, D. L., B. Chan, J. W. M. Bush. 2003. The hydrodynamics of water strider locomotion. *Nature* 424: 663–666.

Hu, D. L., J. W. M. Bush. 2005. Meniscus-climbing insects. *Nature* 437: 733–736.

Kou, J., K. P. Judd, and J. R. Saylor. 2011. The temperature statistics of a surfactant-covered air/water interface during mixed convection heat transfer and evaporation. *International Journal of Heat and Mass Transfer* 54: 3394–3405.

Lee, R. J., and J. R. Saylor. 2010. The effect of a surfactant monolayer on oxygen transfer across an air/water interface during mixed convection. *International Journal of Heat and Mass Transfer* 53: 3405–3413.

Macintyre, F. 1974. The top millimeter of the ocean. *Scientific American* 230: 62–77.

Saylor, J. R. 2001. Determining liquid substrate cleanliness using infrared imaging. *Review of Scientific Instruments* 72: 4408–4414.

Saylor, J. R., and R. A. Handler. 1999. Capillary wave gas exchange in the presence of surfactants, *Experiments in Fluids* 27: 332–338.

Weast, R. C., ed. 1971. *Handbook of Chemistry and Physics, 52nd Edition.* Cleveland, Ohio: Chemical Rubber Company.

Chapter 12

Farnsworth, R. K., E. S. Thompson, E. L. Peck. 1982. Evaporation atlas for the contiguous 48 United States. *NOAA Technical Report NWS 33.*

Howard, B. C. August 12, 2015. Why Did L.A. Drop 96 Million 'Shade Balls' Into Its Water? *National Geographic.* Washington, D.C.: National Geographic Society.

Jones, F. E. 1992. *Evaporation of Water with Emphasis on Applications and Measurements.* Chelsea, Michigan: Lewis Publishers.

Kenny, J. F., N. L. Barber, S. S. Hutson, K. S. Linsley, J. K. Lovelace, M. A. Maupin. 2009. Estimated use of water in the United States in 2005: *U.S. Geological Survey Circular* 1344, 52 p.

La Mer, V. K., ed. 1962. *Retardation of evaporation by monolayers.* New York: Academic Press.

Phillips, R. C., J. R. Saylor, N. B. Kaye, J. M. Gibert. 2016. *Limnology.* 17: 273–289.

Seager, R., A. Tzanova, J. Nakamura. 2009. Drought in the Southeastern United States: causes, variability over the last millennium, and the potential for future hydroclimate change. *Journal of Climate* 22: 5021–5045.

Water Resources Plan, 09. Southern Nevada Water Authority.

Westenburg, C. L., G. A. DeMeo, and D. J. Tanko. 2006. Evaporation from Lake Mead, Arizona and Nevada, 1997–99. *Scientific Investigations Report* 2006-5252. United States Geological Survey.

Chapter 13

Bluestein, G. October 20, 2007. No Backup if Atlanta's Faucets Run Dry. *Washington Post*.

Chappell, B. April 3, 2018. Michigan OKs Nestlé Water Extraction, Despite 80K+ Public Comments Against It. NPR (*https://www.npr.org/sections/thetwo-way/2018/04/03/599207550/michigan-oks-nestl-water-extraction-despite-over-80k-public-comments-against-it*).

Cole, G. A. 1994. *Textbook of Limnology*. Long Grove, Illinois: Waveland Press.

Eagan, D. 2017. *The Death and Life of the Great Lakes*. New York: W. W. Norton & Company.

Fessy, T. April 22, 2016. Why is one of the world's largest lakes disappearing? BBC *News*. (https://www.bbc.com/news/av/36100566).

Golterman, H. L. 1975. *Physiological Limnology: An Approach to the Physiology of Lake Ecosystems*. Amsterdam, The Netherlands: Elsevier Scientific Publishing Company.

Horne, A. J., and C. R. Goldman. 1994. *Limnology*. New York: McGraw-Hill.

Micklin, P. 2007. The Aral Sea disaster. *Annual Review of Earth and Planetary Sciences* 35: 47–72.

National Park Service. April 5, 2017. The third straw. (https://www.nps.gov/lake/learn/the-third-straw.htm).

National Park Service. June 24, 2019. Asian carp overview. (https://www.nps.gov/miss/learn/nature/ascarpover.htm).

Paul, V. G., M. R. Mormile. 2017. A case for the protection of saline and hypersaline environments: a microbiological perspective. FEMS *Microbiology Ecology*. 93: doi:10.1093.

Ross, W. March 31, 2018. Lake Chad: Can the vanishing lake be saved? BBC *News*. (https://www.bbc.com/news/world-africa-43500314).

U. S. Bureau of Reclamation. Lake Mead at Hoover Dam, End of Month Elevation. (https://www.usbr.gov/lc/region/g4000/hourly/mead-elv.html).

Welch, P. S. 1952. *Limnology*. New York: McGraw-Hill Book Company.

Wurtsbaugh, W. A., C. Mille, S. E. Null, J. DeRose, P. Wilcock, M. Hahnenberger, F. Howe, and J. Moore. 2017. Decline of the world's saline lakes. *Nature Geoscience* 10: 816–823.

Chapter 14

Cole, G. A. 1994. *Textbook of Limnology*. Long Grove, Illinois: Waveland Press.

Hutchinson, G. E. 1957. *A Treatise on Limnology, Volume I*. New York: John Wiley & Sons.

Kerr, R. A. 2008. Geologists find vestige of early Earth—maybe world's oldest rock. *Science*. 321: 1755.

Löffler, H. 2004. The Origin of Lake Basins. In P. E. O'Sullivan and C. S. Reynolds, eds. *The Lakes Handbook Volume 1*. Malden, MA: Blackwell Science.

McLennan, S. M. 1992. Continental Crust. In *Encyclopedia of Earth System Science, Volume 1*. Cambridge MA: Academic Press.

O'Neil, J., R. W. Carlson, D. Francis, R. K. Stevenson. 2008. Evidence for Hadean mafic crust. *Science* 321: 1828–1831.

Sandlin, L. 2011. *Wicked River*. New York: Vintage Books.

Taylor, S. R., and S. M. McLennan. 1995. The geochemical evolution of the continental crust. *Review of Geophysics* 33: 241–265.

Taylor, S. R., and S. M. McLennan. 1996. The evolution of continental crust. *Scientific American* 274: 76–81.

United States Geological Survey. The New Madrid Seismic Zone. https://www.usgs.gov/natural-hazards/earthquake-hazards/science/new-madrid-seismic-zone?qt-science_center_objects=0#qt-science_center_objects

ACKNOWLEDGMENTS

I am grateful to the Helene Wurlitzer Foundation for their support in the form of a residency. I thank the Clemson University library staff for their help in locating resources. Dr. Lee Phillips of the University of North Carolina, Greensboro provided valuable insight and information on the Carolina bays. I thank Dr. Martin B. Farley of the University of North Carolina, Pembroke for providing a copy of Raymond T. Kaczorowski's thesis. I thank Dr. Anthony B. Smith for his input and suggestions. I am grateful to all of the people at Timber Press for their support in preparing this book. I thank my agent Laurie Abkemeier for taking a chance on this platformless professor. Most of all I thank my wife, Amy, for her constant encouragement, love, and support.

INDEX